Über die gesetzmäßigen Beziehungen

der

Massenfaktoren

in

normalen Fichtenbeständen.

Von

A. Schiffel.

Separatabdruck aus „Zentralblatt für das gesamte Forstwesen", Heft 5 ex 1903.

Wien 1903.

Wilhelm Frick, k. u. k. Hofbuchhandlung.

Graben 27.

Über die gesetzmäßigen Beziehungen der Massenfaktoren in normalen Fichtenbeständen.

Von A. Schiffel.

Die Untersuchungen über die Beziehungen zwischen den Massenfaktoren gegebener Bestände sind nichts Neues. Die Ursprünge solcher Untersuchungen lassen sich kaum mit Sicherheit verfolgen. Man kann heute nicht konstatieren, wer zuerst Alter, Höhen, Stammzahlen, Formzahlen, Durchmesser, Kreisflächen und Massen eines Bestandes oder einer Reihe von Beständen gleicher Ertrags= fähigkeit untereinander dadurch in Beziehung brachte, daß einer dieser Faktoren als Abszissen=, ein anderer oder ein Produkt dieser Faktoren als Ordinatenachse gewählt und die hierdurch entstehende Kurve graphisch dargestellt wurde; derartige graphische Darstellungen leisteten gute Dienste, um unvermeidliche, natürlich be= dingte Abweichungen von bekannten oder vermuteten Gesetzmäßigkeiten zu elimi= nieren; sie wurden als graphisches Ausgleichs= oder Interpolationsverfahren benutzt und bezeichnet. Neu sind aber die Versuche, in den entstehenden Kurven verschiedener innerer Beschaffenheit einen gleichartigen Typus herauszufinden und diesen Typus allgemein zu verwerten.

Derartige Untersuchungen lassen sich grundsätzlich in zwei Gruppen teilen; eine, welche sich mit den Beziehungen der Massenfaktoren eines und desselben Bestandes in verschiedenen Altersperioden befaßt und hauptsächlich die Darstellung des Entwicklungsganges eines Bestandes nach mathematischen Gesetzen zum Gegen= stande hat, die zweite, welche die Gleichartigkeit der Gesetze, welchen die Massen= faktoren eines Bestandes unterliegen, auch für einen anderen Bestand nachweisen und damit die Herstellung oder Zerlegung der Masse eines gegebenen Bestandes mit möglichst wenig Aufnahmsdaten erreichen will.

In der ersten Gruppe hat Prof. Dr. R. Weber in München grundlegende Arbeiten [1] geliefert und Dr. Gehrhardt [2] interessante Studien gemacht; in die zweite Gruppe gehören die Arbeiten des k. k. Forstrates R. Kopezky [3] und des Prof. Ludwig Fekete [4] in Schemnitz.

---

[1] Allgemeine Forst= und Jagdzeitung: „Die Gesetzmäßigkeit im Zuwachsgange der Waldbestände" 1903. „Über die Gesetzmäßigkeit im Zuwachsgange einiger Holzarten auf Grund neuerer Ertragstafeln". 1. „Das Höhenwachstum" 1895. 2. „Das Dickenwachstum" 1898. „Über die mathematischen Beziehungen zwischen dem arithmetischen Mittelstamm und der Bestandesmasse" 1899.

[2] „Die theoretische und praktische Bedeutung des arithmetischen Mittelstammes". Meiningen 1901.

[3] Österr. Vierteljahrsschrift für Forstwesen. „Die Flächestufen und deren Anwendung in der Holzmeßkunde" 1903.

[4] Erdészeti kisérletek. Schemnitz 1902, Heft Nr. 3.

Die nachfolgende Betrachtung soll sich nur auf einen Spezialteil der zweiten Gruppe, nämlich auf die gesetzmäßigen Beziehungen der Massenfaktoren in normalen Fichtenbeständen beschränken und hauptsächlich die praktische Verwertbarkeit der gefundenen Resultate im Auge behalten.

Kopezkys Hauptsätze sind folgende:

1. Die mittleren Baummassen der Stärkestufen eines gegebenen Bestandes bilden eine gerade Linie, wenn sie als Ordinaten der zugehörigen Grundflächen (Abzissen) aufgetragen werden.

2. Die mittleren Formhöhen (Baumformzahl $\times$ Höhe) der einzelnen Stärkestufen ergeben als Funktion der Grundflächen eine gleichseitige Hyperbel.

3. Die Produkte g h (Grundfläche $\times$ Höhe) und g f_b (Grundfläche $\times$ Baumformzahl) bilden als Funktionen ihrer Grundflächen je für sich eine Gerade.

4. Die mittleren Höhen (h) und die mittleren Baumformzahlen (f_b) der einzelnen Stärkestufen verlaufen als Funktionen der Grundflächen je für sich in der Form einer gleichseitigen Hyperbel.

Alle diese Kurven sind in ihrem Verlaufe bestimmt, wenn zwei Punkte derselben durch ihre Koordinaten gegeben sind. Die praktische Bedeutung dieser Gesetze liegt also darin, daß in einem gegebenen Bestande bloß die Masse und Formhöhen zweier zweckmäßig gewählter Stärkestufen zu erheben sind, um auf Grund dieser Erhebungen die Massen und Formhöhen jeder beliebigen Stärkestufe bestimmen zu können.

Kopezky bezieht seine Gesetze alle auf die Grundflächen als Abzissenachse. Dieser Umstand verlangt aber durchaus nicht, daß die Messung der Bäume nach Flächestufen und nicht nach Durchmessern erfolgen müsse. Das erste Gesetz Kopezky lautet: y = a x — b, wobei y = v, x = g zu setzen ist. a und b sind zwei für den gegebenen Bestand konstante Größen. Diese Form genügt vollständig, um das Kopezkysche Massen= und Formhöhengesetz abzuleiten, ohne daß hierzu das Beiwerk von Grundzahl, Zuwachskoeffizient, Zuwachscharakteristik, namentlich aber die Kluppierung nach Flächenstufen erforderlich wäre. Sind a und b aus zwei Massen= und Grundflächenerhebungen bestimmt, so kann man auch die Masse einer jeden beliebigen Stärkestufe daraus berechnen, weil die Kopezkyschen Flächenstufen Kreisflächenstufen sind und deshalb $g = \dfrac{d^2 \pi}{4}$ ist.

Ich habe das Kopezkysche Massengesetz in Fichtenbeständen untersucht und bezüglich der Anwendung auf Baumholzmassen sehr gute Übereinstimmung mit der Wirklichkeit gefunden. Die wissenschaftliche Bedeutung dieses Gesetzes — immer in der Anwendung auf Baumholz gedacht — ist gar nicht zu bestreiten. Der praktischen Anwendung stehen einige Schwierigkeiten im Wege. Diese sind:

1. Das Gesetz gilt nur für Baummassen. Der Praxis genügt jedoch die Gesamtmasse nicht und es müßte erst eine Teilung in Schaft= oder Derbholz und Reisholz erfolgen, welche auch bei Anwendung entsprechender Hilfstafeln an Schwierigkeiten stößt und die Fehlerquellen vermehrt.

2. Der Gebrauch erfordert die vollständige Auskluppierung des Bestandes und die Bestimmung der Baummassen, beziehungsweise Formhöhen zweier Stärkestufen. Will man sich vor größeren Fehlern hüten, so wird es unerläßlich sein die Probestämme in mehreren Exemplaren zu fällen. Diese Erfordernisse genügen bei entsprechender Stammklassenbildung und Probestammverteilung, um nicht nur die Baummassen, sondern auch Schaft= und Derbholzmassen, Höhen= und Formzahlen direkt zu bestimmen.

3. Die Praxis bedarf bei Massenaufnahmen von Beständen auch der Höhen, weil diese zur Sortimentsbildung und Bewertung erforderlich sind. Die Kopezkysche Massenformel läßt aber eine Trennung der durch

$$g\,h\,f = a\,g - b$$

$$h\,f = a - \frac{b}{g},$$

oder analytisch ausgedrückt:

$$y = a - \frac{b}{x} \quad \text{(Hyperbel)},$$

bestimmten Formhöhen nicht zu. Hierzu müßten seine Spezialformeln herangezogen werden, deren Brauchbarkeit jedoch noch nicht genügend erhärtet ist. Zur Bestimmung der Höhe wäre sonach die Formzahl erforderlich und umgekehrt.

Aus diesen Einwänden, welche sich lediglich auf den praktischen Gebrauch beziehen und die wissenschaftliche Bedeutung der Forschungsresultate nicht tangieren, lassen sich als Erfordernisse praktisch verwertbarer Gesetze folgende Grundsätze aufstellen:

1. Die im Bestande zu erhebenden Daten dürfen einen Umfang nicht erreichen, welcher erforderlich ist, um auch ohne Anwendung von theoretischen Hilfsmitteln eine entsprechende Massenermittlung und die Zerlegung der Masse zu Sortimenten vornehmen zu können.

2. Die aufzustellenden Gesetze sollen daher die Ermittlung der Derbholz- oder Schaftmassen und zugehörigen Höhen — wenn auch nicht für jede einzelne Stärkestufe — so doch für eine entsprechende, den Verhältnissen angepaßte Anzahl von Stärkeklassen ohne Fällung von Probestämmen in diesen Klassen ermöglichen.

Was zunächst die Bedingung 1 anbelangt, so ist ohne weiteres einzusehen, daß die Massenermittlung nach dem arithmetischen Mittelstamme die einfachste wenn auch nicht sicherste Methode der Bestandesaufnahme ist. Es handelt sich also darum, mit Hilfe der Daten: Stammzahl, Höhe, Formzahl und Mittelstammdurchmesser, welche bei der Massenaufnahme nach dem Mittelstamme gewonnen werden, Gesetze ausfindig zu machen, die dennoch eine Zerlegung der Masse des Bestandes in Sortimente nach Stärkestufen oder Stammklassen gestatten.

Gelingt dies, dann erweitert sich die praktische Anwendbarkeit solcher Gesetze ganz erheblich und läßt sich auf ein Gebiet erstrecken, welches als dankbarstes Objekt dafür anzusehen ist.

Die modernen Ertragstafeln enthalten nämlich alle diese als Minimum bezeichneten Daten. Ein oft und schwer empfundener Übelstand bei der Anwendung von Ertragstafeln, insbesondere bei Waldwertberechnungen, Rentabilitäts-, Wertzuwachsprozent und Umtriebszeitkalkulationen ist es, daß die Bewertung der Massen entweder nach dem Mittelstamme oder im Anhalte an sonstige unsichere Daten (Schlagergebnisse) vorgenommen werden muß. Es ist ein ganz berechtigter Einwand, den man gegen solche Wertsermittlungen der Haubarkeitsnutzung geltend macht, daß die Bewertung der Masseneinheit nach finanziellen Ertragstafeln mehr weniger eine willkürliche ist und einer positiven Grundlage entbehrt. Wird daher eine annehmbare Methode der Zerlegung der Bestandesmassen mit Hilfe der in der Ertragstafel enthaltenen Angaben: Stammzahl, Mittelstammdurchmesser, Höhe und Formzahl gefunden, so ist auch das Mittel gegeben, die gewünschte Sortierung zur Bewertung vorzunehmen.

Einen wesentlichen Beitrag zur Erreichung dieses Zieles hat Oberforstrat Fekete geleistet. Bekannt ist die Erscheinung, daß sich in einem regelmäßigen Bestande die Anzahl der in den einzelnen Stärkestufen vorhandenen Stämme mit einer gewissen Gesetzmäßigkeit gruppiert;[1] bekannt ist auch das Weisesche Gesetz: Der arithmetische Mittelstamm fällt in diejenige Stärkestufe, zu der man gelangt,

---

[1] Vgl. Loreys Handbuch der Forstwissenschaft 1903. Holzmeßkunde, v. Guttenberg, II. Bd., S. 228.

wenn man 40 Prozent der Stammzahl vom stärksten Stamme angefangen gegen die schwächeren fortschreitend abzählt.

Fekete hat eine größere Anzahl von Fichtenbeständen in bezug auf die Gesetzmäßigkeit des Verhaltens der Durchmesser zu dem ihnen zugehörigen Prozentanteile der Stammzahl in der Weise untersucht, daß er die Durchmesser von 10 zu 10 Prozent der Stammzahl bestimmte. Hierbei ergab sich, daß die Bestände gleichen Mittendurchmessers bei gleichen Prozenten auch annähernd gleiche Durchmesser aufwiesen. Durch entsprechende Mittelbildung gelangte er zu der nachstehend reproduzierten Tabelle, welche nach Abstufungen von je 5 cm Mittendurchmesser die Durchmesser der Stammzahlenprozente in Abstufungen von 10 Prozent angeben.

Durchmesser des Mittelstammes	Bei den Prozenten der Stammzahl										
	1	10	20	30	40	50	60	70	80	90	100
cm	beträgt der Durchmesser cm										
10	5·4	7·1	7·7	8·1	8·5	9·1	9·7	10·5	11·5	12·8	19·5
15	8·2	10·5	11·5	12·4	13·1	14·0	14·9	16·0	17·5	19·2	26·5
20	11·0	13·9	15·4	16·6	17·7	18·8	20·1	21·5	23·3	25·8	33·4
25	13·8	17·3	19·3	20·8	22·3	23·7	25·2	27·0	29·2	32·0	40·3
30	16·6	20·7	23·1	25·1	26·8	28·6	30·3	32·5	35·1	38·3	47·2
35	19·4	24·1	27·0	29·3	31·4	33·5	35·5	37·9	41·0	44·8	54·1
40	22·2	27·5	30·9	33·6	36·0	38·4	40·7	43·4	46·8	51·1	61·0
45	25·0	30·9	34·7	37·9	40·6	43·3	45·9	48·9	52·8	57·5	67·9
50	27·8	34·3	38·7	42·1	45·2	48·2	51·0	54·4	58·6	63·9	74·7

Fekete hat die Ergebnisse dieser Untersuchung auch graphisch dargestellt, indem er die Stammzahlenprozente als Abszissen, die zugehörigen Durchmesser als Ordinatenachse benutzte. Hierbei ergaben sich Kurven, die zuerst konvex verlaufen und mit einem Wendepunkte in einem konkaven Teil endigen. Durch entsprechende Interpolation wäre es ein Leichtes, diese Kurven nach beliebigen Abstufungen graphisch zu ergänzen, womit man die Daten gewänne, um bei gegebenem Durchmesser des Mittelstammes die Durchmesser abzulesen, welche ein beliebiges Prozent der Stammzahl besitzt. Der hervorragende Wert dieser Forschung ist also: Bei gegebener Stammzahl und bekanntem Mittelstammdurchmesser ist bei prozentueller Verteilung der Stammzahl der Durchmesser eines jeden beliebigen Prozentes der Stammzahl bestimmt. Prof. Fekete hat eine praktische Verwertung seiner Forschungsresultate vorläufig nicht ins Auge gefaßt und nach einem mathematischen Ausdrucke seiner aus Fichtenbeständen guter Bonität abgeleiteten Kurven nicht gesucht. Es ist jedoch leicht einzusehen, daß die Feketeschen Kurven für die früher bezeichneten praktischen Ziele ausgezeichnete Dienste zu leisten im stande sind, ja, die wichtigste Voraussetzung für die Erreichung dieser Ziele bilden, wenn sie eine allgemeine Giltigkeit besitzen.

Ich habe die Feketeschen Durchmesserkurven an einer größeren Anzahl von Fichtenbeständen nachgeprüft und gefunden, daß sie für Fichten-Normalbestände d. h. für gleichalterige, im vollen Schlusse bei mäßiger Durchforstung erwachsene Bestände jeder Bonität mehr als hinreichende annähernde Giltigkeit besitzen, um sie der beabsichtigten praktischen Verwertung dienstbar zu machen, daß sie auch noch bei größeren Altersunterschieden und selbst bei Fichtenbeständen, welche mit anderen Nadelhölzern gemischt sind, brauchbar sind, wenn in den Höhen der bestandbildenden Holzarten keine wesentlichen Unterschiede bestehen und im Bestande keine, das natürliche Wachstum wesentlich alterierenden Holzentnahmen vorgenommen wurden. Durchforstungen, welche sich auf alle Stärkestufen gleichmäßig verteilen, bilden kein Hindernis für ihre Anwendung; dagegen passen die Kurven nach plötzlicher Entfernung vorwiegend schwacher oder vorwiegend starker Stämme und für ausgesprochen unregelmäßige Bestände (Alter und Schlußform) nicht. Die Giltigkeit der Feketeschen Kurven erstreckt sich also auf ein weites Gebiet. In der folgenden Zusammenstellung ist eine Reihe von Fichtenbeständen dargestellt, welche derart ausgewählt sind, daß sie als Demonstrationsobjekte für die nachfolgenden Ausführungen dienen können. Die Vorführung von Material in breiterem Rahmen erscheint unnötig, da jeder Forstwirt, der es mit Fichtenbeständen zu tun hat, über Probeflächen verfügt, an denen er die Resultate der folgenden Studienergebnisse prüfen kann.

Zunächst erschien es wünschenswert, einen möglichst allgemeinen mathematischen Ausdruck für die Feketeschen Kurven, d. i. ihr Gesetz aufzustellen, um die theoretischen Anhaltspunkte zur Bestimmung der Mittelstämme beliebig gewählter Stammklassen zu gewinnen. In der Fig. 1 sind die Kurven für die Mittelstämme von 15, 25, 35 und 45 cm Durchmesser, wie sie sich nach Fekete ergeben, dargestellt.

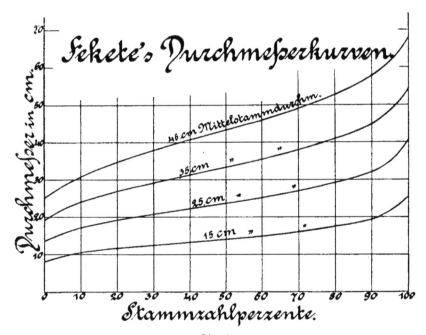

Fig. 1.

Zur näherungsweisen Bestimmung dieser Art von Kurven reicht der Typus

$$y = a + bx + cx^2 + dx^3$$

aus. Es sind also Potenzkurven dritter Ordnung, durch einen Wendepunkt in zwei unsymmetrische Äste geteilt; sie sind endlich begrenzt und liegen zwischen $x = 0$ und $x = 100$. Berechnet man die Parameter a, b, c und d nach obiger Gleichung, so wird man finden, daß sie für jede Kurve verschieden sind. Die entstehenden Gleichungen wären nichts anderes als ein mathematischer Ausdruck jeder einzelnen

Bestandescharakteristik					Bei den		
Alter	Stammzahl	Höhe	Formzahl	Mitteldurch-messer	beträgt der Durchmesser und die Höhe	0	10
45	1472	19·0	0·52	18·0	d wirklich d berechnet h wirklich h berechnet	9·0 10·0 13·2 12·9	11· 12· 15· 15·
52	1344	21·0	0·53	19·1	d wirklich d berechnet h wirklich h berechnet	8·5 10·6 13·2 14·3	11· 13· 15· 16·
40	1960	14·1	0·54	12·2	d wirklich d berechnet h wirklich h berechnet	6·0 6·8 9·2 9·6	8· 8· 10· 11·
53	1420	20·5	0·55	18·9	d wirklich d berechnet h wirklich h berechnet	10·0 10·5 14·7 13·9	13· 13· 16· 16·
77	1270	24·2	0·515	22·1	d wirklich d berechnet h wirklich h berechnet	11·0 12·3 16·2 16·4	15· 15· 19· 19·
104	777	23·0	0·52	27·2	d wirklich d berechnet h wirklich h berechnet	15·0 15·0 16·4 15·6	19· 18· 18· 18·
55	2028	15·3	0·54	13·2	d wirklich d berechnet h wirklich h berechnet	7·0 7·3 10·8 10·4	9· 9· 12· 12·
52	1585	17·3	0·52	19·1	d wirklich d berechnet h wirklich h berechnet	7·3 10·6 11·1 11·7	11· 13· 12· 13·
78	2000	15·8	0·53	14·5	d wirklich d berechnet h wirklich h berechnet	7·1 8·0 9·5 10·7	9· 10· 11· 12·
86	955	31·7	0·48	30·0	d wirklich d berechnet h wirklich h berechnet	16·2 16·6 23·0 21·6	20· 20· 26· 25·
94	1066	27·1	0·50	27·6	d wirklich d berechnet h wirklich h berechnet	15·1 15·3 19·6 18·4	18· 19· 22· 21·
94	815	23·3	0·49	29·1	d wirklich d berechnet h wirklich h berechnet	15·0 16·1 16·2 15·8	20· 20· 19· 18·

Stammzahlenprozenten

20	30	40	50	60	70	80	90	100
13·6	14·5	16·5	17·5	18·4	19·5	21·4	24·5	29·0
13·9	15·1	16·1	17·2	18·2	19·4	21·1	23·9	28·9
16·9	17·5	18·4	18·8	19·1	19·5	19·8	20·3	20·6
16·4	17·3	18·0	18·6	19·1	19·8	20·0	20·7	21·7
13·9	15·8	17·2	18·5	19·8	21·3	22·5	24·8	29·6
14·7	15·9	17·1	18·2	19·3	20·6	22·3	24·4	29·8
17·7	19·3	20·0	20·7	21·3	21·6	21·9	22·2	22·4
18·2	19·1	19·9	20·5	21·8	21·1	22·1	22·8	23·9
9·4	10·0	10·7	11·6	12·3	13·2	14·4	15·7	20·1
9·4	10·2	10·9	11·6	12·3	13·2	14·3	15·6	19·0
12·2	12·7	13·3	13·8	14·2	14·6	15·4	15·9	17·0
12·2	12·8	13·4	13·8	14·2	14·7	14·9	15·4	16·1
14·7	15·6	16·8	18·0	19·4	20·4	22·6	25·4	31·0
14·6	15·8	16·9	18·0	19·1	20·4	22·1	24·2	29·5
17·8	18·4	19·3	20·0	20·5	21·2	21·7	22·4	23·0
17·8	18·7	19·4	20·0	20·6	21·3	21·6	22·3	23·4
16·7	18·3	19·4	20·6	22·4	23·6	25·5	29·0	36·0
17·0	18·5	19·8	21·1	22·3	23·8	25·9	28·3	34·5
20·7	22·2	22·8	23·4	24·4	24·7	25·6	26·7	27·7
20·9	22·0	22·9	23·7	24·3	25·1	25·5	26·4	27·6
21·7	23·0	24·3	25·8	27·0	29·0	31·7	35·0	46·3
21·0	22·8	24·3	26·0	27·4	29·4	31·8	34·8	42·4
20·3	21·2	21·8	22·4	22·9	23·7	24·6	25·8	27·4
19·9	21·0	21·8	22·5	23·1	23·9	24·2	25·1	26·2
10·4	11·0	11·7	12·4	13·0	14·0	15·3	17·1	22·2
10·2	11·0	11·8	12·6	13·3	14·3	15·4	16·9	20·6
13·3	13·7	14·3	14·8	15·2	15·7	16·2	16·6	17·5
13·2	13·9	14·5	14·9	15·4	15·9	16·1	16·7	17·4
13·7	15·1	16·4	18·0	19·5	20·8	22·4	25·0	32·6
14·7	16·0	17·1	18·2	19·2	20·6	22·3	24·4	29·8
14·5	15·6	16·4	17·0	17·3	17·6	17·8	18·2	18·5
15·0	15·8	16·4	16·9	17·4	18·0	18·2	18·9	19·7
10·5	11·5	12·6	13·6	15·0	16·0	17·5	20·4	24·3
11·2	12·1	13·0	13·8	14·6	15·7	16·9	18·6	22·6
12·9	13·8	14·6	15·3	16·1	16·6	17·4	18·5	19·8
13·7	14·4	15·0	15·4	15·9	16·4	16·6	17·2	18·0
22·7	24·5	26·6	28·0	30·3	33·0	35·5	39·0	46·3
23·1	25·1	26·8	28·6	30·3	32·4	35·1	38·4	46·8
28·1	28·8	29·7	30·2	30·8	31·5	32·0	32·6	33·0
27·5	28·9	30·0	31·0	31·9	32·9	33·4	34·5	36·1
21·4	23·5	25·0	26·5	28·4	30·0	32·4	35·4	45·1
21·3	23·1	24·7	26·3	27·9	29·8	32·3	35·3	43·1
24·3	25·4	26·2	26·8	27·5	28·1	28·6	29·8	31·8
23·4	24·5	25·7	26·5	27·2	28·2	28·5	29·5	30·8
22·0	24·1	25·9	28·2	29·6	31·5	33·8	38·0	47·0
22·4	24·4	26·0	27·8	29·4	31·4	34·0	37·3	45·3
20·2	21·2	22·2	23·0	23·5	24·0	24·7	25·3	27·0
20·2	21·2	22·1	22·8	23·4	24·2	24·5	25·4	26·5

Kurve. Da weiters diese Kurven durch nichts als solche gekennzeichnet sind, daß sie zu einem bestimmten Mittelstammdurchmesser gehören als durch die Aufschrift, hätte der mathematische Ausdruck vor dem graphischen keinerlei Vorzug und das Gesetz würde nach wie vor lauten:

Bestände, welche bei einem gleichen Stammzahlenprozente die gleichen Durchmesser aufweisen, haben auch bei allen übrigen gleichen Stammzahlenprozenten die gleichen Durchmesser. Es ist also in dem Gesetze eine spezielle Beziehung zum Mittendurchmesser nur dann gegeben, wenn man eben den Mittendurchmesser zum Vergleiche wählt.

1. Die Durchmesserkurve.

Um eine allgemeine mathematische Beziehung zwischen Stammzahlenprozenten und dem Durchmesser herzustellen, muß zunächst untersucht werden, ob bei allen Kurven die gleichen Beziehungen zwischen den Durchmessern gleicher Prozente, oder, um gleich auf das gewählte Ziel loszusteuern, ob bei allen Kurven die gleichen Beziehungen zwischen Mittelstammdurchmesser und den Durchmessern bei gleichen Stammzahlenprozenten bestehen, d. h. ob alle Kurven einem einzigen gleichen Gesetze folgen, in welchem die Stammzahlenprozente die unabhängig Veränderliche, die Durchmesserrelation die abhängig Veränderliche ist.

Bringen wir den Durchmesser des Mittelstammes d in ein Verhältnis zu einem anderen beliebigen Durchmesser d_n des Bestandes, so erhalten wir in $R_d = d_n : d$ eine Reduktionszahl, mit welcher der Mittelstammdurchmesser zu multiplizieren ist, um den Durchmesser des bei n Prozent der Stammzahl liegenden Stammes zu finden. Wird die Reduktionszahl mit 100 multipliziert, so gibt sie den Durchmesser d_n in Prozenten von d an. Die Feketesche Tabelle auf Seite 4 ergibt nach dieser Behandlung folgende Resultate:

Mittelstamm-durchmesser d	Bei den Stammzahlenprozenten										
	0	10	20	30	40	50	60	70	80	90	100
	beträgt der Durchmesser Hundertelteile (Prozente) des Mittendurchmessers										
10	54·0	71·0	77·0	81·0	85·0	91·0	97·0	105	115	128	195
15	54·7	70·0	76·6	82·7	87·1	93·3	99·3	107	117	128	177
20	55·0	69·5	77·0	83·0	88·5	94·0	100·5	107	117	129	167
25	55·2	69·2	77·2	83·2	89·2	94·8	101·0	108	117	128	161
30	55·3	69·0	77·1	83·8	89·3	95·3	101·0	108	117	128	157
35	55·5	68·9	77·1	83·8	89·7	95·8	101·0	108	117	128	155
40	55·5	68·7	77·2	84·0	90·0	96·0	102·0	108	117	128	152
45	55·7	68·7	77·1	84·2	90·2	96·2	102·0	108	117	128	151
50	55·6	68·6	77·4	84·2	90·0	96·4	102·0	109	117	128	149
Im Mittel	55·5	68·9	77·1	83·7	89·5	95·5	101	108	117	128·1	156

Aus dieser Zusammenstellung ist zu ersehen, daß von 20 cm Mittelstammdurchmesser aufwärts die Mittelstammdurchmesserprozente bei gleichen Stammzahlenprozenten sehr angenähert konstant bleiben, eine Ausnahme hiervon besteht nur bei dem Stammzahlenprozent 100, das ist be den letzten Ausläufern der Stärkestufen. Da jedoch die Durchmesserprozente schon bei 90 Prozent auffallend gleich sind, besteht kein Anlaß, diese Unregelmäßigkeit bei 100 Prozent zu berücksichtigen. Für die praktische Verwertung des nunmehr in anderer Form dargestellten Gesetzes kommen Mittelstammdurchmesser unter 20 cm kaum in Betracht, weshalb ich auch die ersten beiden Reihen bei der Mittelbildung nicht berücksichtigte. Übrigens ist zu ersehen, daß auch für geringere Durchmesser als 20 cm die Mittelwerte noch annähernde Geltung haben.

Durch diese einfache Operation ist also erreicht, daß wir anstatt der vielen für jeden Mittelstammdurchmesser verschiedenen Kurven eine einzige gewinnen, in welcher dem absoluten Durchmesser irgend eines Stammzahlenprozentes keine Rolle mehr zufällt; das Gesetz lautet nunmehr in bemerkenswert allgemeinerer Form: In jedem normalen Fichtenbestande entspricht einem bestimmten Stammzahlenprozente ein gleiches Prozent des Mittelstammdurchmessers. Wäre beispielsweise bei einem Bestande der Mittelstammdurchmesser 20 cm, bei einem anderen 40 cm und wird der Durchmesser für 30 Prozent der Stammzahl gesucht, so ist in beiden Fällen die Reduktionszahl die gleiche, nämlich 0·837 und wir finden die gesuchten Durchmesser 20 × 0·837 = 16·7 cm, beziehungsweise 40 × 0·837 = 33·5 cm.

Um einen mathematischen Ausdruck für dieses Gesetz zu finden, wählen wir die Stammzahlenprozente als Abszissenachse, die Durchmesserreduktionszahlen als Ordinatenachse und tragen die von 10 zu 10 Prozent steigenden Werte der Reduktionszahlen als Ordinaten der zugehörigen Abszissen auf. Verbinden wir die so entstehenden Punkte miteinander, so erhalten wir die in der Fig. 2 dargestellte Kurve.

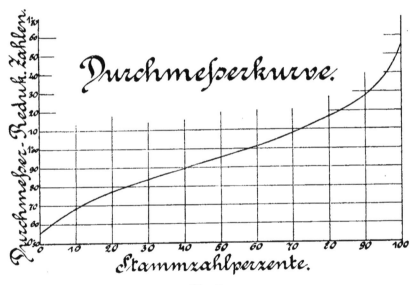

Fig. 2.

Diese Kurve hat eine den Feketeschen Kurven, aus denen sie abgeleitet ist, ähnliche Form. Der zugehörige Kurventypus, welcher diese Linie annähernd umschreibt, ist auch hier:

$$y = a + b\,x + c\,x^2 + d\,x^3.$$

In dieser Gleichung läßt sich a eliminieren, wenn wir die x-Achse um das Stück a = 0·555 parallel zu sich selbst verschieben. Es bleibt nunmehr die Bestimmung der Konstanten b, c und d übrig, welche mit Hilfe dreier zweckmäßig gewählter Gleichungen erfolgen kann. Nach mehrfachen Versuchen zeigte sich, daß die gerechnete Kurve, welche in Fig. 2 dargestellt ist, der aus den Reduktionszahlen direkt konstruierten am nächsten liegt, wenn die Werte der Abfziffen 15, 80 und 95 Prozent der Berechnung der Konstanten zugrunde gelegt werden. Die auf diesem Wege erhaltene Gleichung ist:

$$R_d = 0{\cdot}555 + 0{\cdot}0144\,N_p - 0{\cdot}000209\,N_p^2 - 0{\cdot}00000157\,N_p^3 \ldots 1.$$

Wird die Berechnung der Mittelstammdurchmesser-Reduktionszahlen nach dieser Formel durchgeführt, dann erhält man für die Stammzahlenprozente:

| 10 | 20 | 30 | 40 | 50 | 60 | 70 | 80 | 90 | 100 |

die Reduktionszahlen für den Mittelstammdurchmesser:

| 0·680 | 0·771 | 0·841 | 0·898 | 0·948 | 1·006 | 1·078 | 1·173 | 1·302 | 1·475 |

Diese mit der Formel 1 gerechneten Resultate differieren nur um ein Geringes gegen die wirklichen Mittelwerte; größere Abweichungen kommen nur bei 90 und 100 Prozent vor. Da jedoch die Werte bei 80 und 95 Prozent übereinstimmen, so hat diese Abweichung keine praktisch ins Gewicht fallenden Folgen.

Mit Hilfe der Formel 1 ist man demnach in der Lage, ganz allgemein für einen beliebigen Bestand, dessen Mittendurchmesser bekannt ist, die Durchmesser beliebiger Stammzahlenprozente zu berechnen. Bei der praktischen Verwertung dieses Gesetzes kommt es jedoch nicht darauf an, für jedes Stammzahlenprozent den Durchmesser zu finden, sondern man wird sich damit begnügen können, die mittleren Durchmesser entsprechend gewählter Stärkeklassen zu bestimmen, weil die Kubierung jeder Stammzahlenprozentstufe für sich eine unnötige und in Anbetracht der Methode als Näherungsverfahren auch eine undankbare Aufgabe wäre. Zur Zerlegung der Bestandesmasse in Sortimente reichen fünf Stammklassen in der Regel aus. Es wird sich also darum handeln, die Stammklassenmittelstämme von höchstens fünf Stammklassen zu entwickeln. Wären die mittleren Durchmesser der Stammklassen einfache arithmetische Mittel aller in der Stammklasse vereinigten Durchmesser, so wäre diese Aufgabe sehr leicht auf rechnerischem Wege in der Weise zu lösen, daß man für das in einer Stärkeklasse vereinigte Stammzahlenprozent die Fläche bestimmt — was durch Differentiation und Integration

nach $F = \int y\,dx$ leicht tunlich wäre — und aus $\dfrac{F}{N_p}$ den mittleren Durchmesser

berechnet. Da wir jedoch unter Mittelstamm denjenigen verstehen, welcher die mittlere Grundfläche aufweist, müssen wir einen anderen Weg wählen. Konstruieren wir uns einen Idealbestand mit beliebig gewählter Stammzahl und beliebigem Mittelstammdurchmesser und berechnen wir für jedes Stammzahlenprozent den zugehörigen Durchmesser nach Formel 1, so haben wir alle Daten, um für beliebig gewählte Stärkeklassen die mittlere Grundfläche und den zugehörigen Durchmesser zu bestimmen, in ganz analoger Weise, wie dies bei Bestandesaufnahmen geschieht. Die Resultate dieses Idealbestandes haben natürlich als Mittelwerte für alle Normalbestände Geltung. Auf diesem Wege ergab sich:

A. Stärkeklassen mit gleicher Stammzahl in jeder Stärkeklasse:

a) 5 Stärkeklassen, jede mit 20 Prozent der Stammzahl dotiert.

Stärkeklasse I   II   III   IV   V

Der Mittelstammdurchmesser liegt bei

| 10 | 30 | 51 | 71 | 91 Prozent der Stammzahl. |

Der Durchmesser des Mittelstammes der Stärkeklasse beträgt

| 0·685 | 0·842 | 0·963 | 1·09 | 1·36 des Bestandesmittelstammdurchmessers. |

b) 4 Stärkeklassen, jede mit 25 Prozent der Stammzahl dotiert.

Stärkeklasse: I   II   III   IV

Der Mittelstammdurchmesser liegt bei

| 13 | 44 | 60 | 89 Prozent der Stammzahl. |

Der Durchmesser des Mittelstammes der Stärkeklasse beträgt
    0·710 | 0·917 | 1·01 | 1·287 des Bestandesmittelstammdurchmessers.
c) 3 Stärkeklassen, jede mit ⅓ der Stammzahl dotiert.
    Stärkeklasse I   II   III
Der Mittelstammdurchmesser liegt bei
        16 | 51 | 82 Prozent der Stammzahl.
Der Durchmesser des Mittelstammes der Stammklasse beträgt
        0·738 | 0·963 | 1·196 des Bestandesmittelstammdurchmessers.

B. Stärkeklassen mit gleicher Grundfläche in jeder Stärkeklasse.

a) 5 Stärkeklassen, jede mit 20 Prozent der Grundfläche dotiert.
Stärkeklasse I   II   III   IV   V
    Der Mittelstammdurchmesser liegt bei
        17 | 48 | 68 | 84 | 95 Prozent der Stammzahl.
    Der Durchmesser des Mittelstammes der Stärkeklasse beträgt
    0·746 | 0·938 | 1·060 | 1·220 | 1·380 des Bestandesmittelstammdurchmessers.
    Die Stärkeklasse umfaßt
    35·5 | 22·8 | 17·7 | 13·5 | 10·5 Prozent der Stammzahl.
b) 4 Stärkeklassen, jede mit 25 Prozent der Grundfläche dotiert.
Stärkeklasse I   II   III   IV
    Der Mittelstammdurchmesser liegt bei
        20 | 60 | 73 | 94 Prozent der Stammzahl.
    Der Durchmesser des Mittelstammes der Stärkeklasse beträgt
        0·771 | 1·006 | 1·103 | 1·366 des Bestandesmittelstammdurchmessers.
    Die Stärkeklasse umfaßt
        41·3 | 24·7 | 20·5 | 13·5 Prozent der Stammzahl.
c) 3 Stärkeklassen, jede mit ⅓ der Grundfläche dotiert.
    Stärkeklasse I   II   III
    Der Mittelstammdurchmesser liegt bei
        25 | 67 | 92 Prozent der Stammzahl.
    Der Durchmesser des Mittelstammes der Stärkeklasse beträgt
        0·814 | 1·055 | 1·333 des Bestandesmittelstammdurchmessers.
    Die Stärkeklasse umfaßt
        51 | 30 | 19 Prozent der Stammzahl.

Mit diesen Daten reicht man in der Praxis vollkommen aus, um die Zerlegung der Stammzahl in Stärkeklassen vornehmen und in jeder derselben den Durchmesser des Mittelstammes bestimmen zu können, wenn die Stammzahl und der Durchmesser des Bestandesmittelstammes bekannt sind.

Hiermit ist jedoch erst die Grundfläche bestimmt und es erübrigt uns zur Massenbildung noch die Ermittlung der Höhen= und Formzahlen.

2. Die Höhenkurve.

Die zweifellos nachgewiesenen gesetzmäßigen Beziehungen zwischen Stammzahlenprozenten und ihren Durchmessern lassen vermuten, daß eine solche Gesetzmäßigkeit auch zwischen Stammzahlenprozenten einerseits, Höhen= und Formzahlen anderseits bestehe.

Allerdings ist bei letzteren der Nachweis insofern schwieriger, als die Höhen= und Formzahlen der einzelnen Stammzahlenprozente der Messung, beziehungsweise Ermittlung nicht so leicht und sicher zugänglich sind, wie die Durchmesser. Der forstlichen Versuchsanstalt in Mariabrunn steht jedoch eine, wenn auch beschränkte Anzahl von Versuchsflächen zur Verfügung, in welchen die Höhen verschiedener Stärkestufen aus einer größeren Anzahl von Probestämmen mit Sicherheit ermittelt werden konnten. Diese Höhen wurden graphisch ausgeglichen, so daß die jeder Stärkestufe zukommende, durchschnittliche Höhe abgelesen werden konnte. Aus diesen Versuchsflächen wurden die auf S. 6 angeführten, mit tunlichster Rücksicht darauf ausgewählt, daß verschiedene Bonitäten und Massenzusammensetzungen zur Darstellung gelangen.

Werden die in der bezogenen Tabelle angeführten, den verschiedenen Stammzahlenprozenten zukommenden Höhen in der gleichen Weise wie die Durchmesser in Prozenten der Mittelstammhöhe ausgedrückt, so ergibt sich folgende Tabelle:

Mittlere Bestandeshöhe h	Bei den Stammzahlenprozenten										
	0	10	20	30	40	50	60	70	80	90	100
	beträgt die Höhe Hundertelteile (Prozente) der Mittelstammhöhe										
19·0	69 7	81·1	89·0	93·7	96·9	99·1	100·3	102·8	104·2	106·9	108·5
21·0	63·3	73·3	84·3	92·0	95·2	98·6	101·5	10ʻ9	104·5	105·8	106·8
14·1	65·4	76·6	86·4	90·1	94·3	97·9	100·8	103·5	109·3	112·5	121·0
20·5	68·7	79·1	86·8	89·8	94·1	97·6	100·0	103·5	105·8	109·3	112·3
24·2	67·0	78·9	85·6	91·8	94·2	96·7	100·8	102·7	105·5	110·0	114·5
23·0	71·4	81·7	88·3	92·2	94·8	97·4	99·8	103·0	107·0	112·1	119·1
15·3	70·3	82·4	86·9	89·6	93·4	96·8	99·4	103·6	105·9	108·6	114·5
17·3	64·4	74·1	83·9	90 2	94·8	98·3	99·4	101·8	102·9	105·4	107·0
15·8	60·0	70·3	81·7	87·4	92·5	96·8	102·1	105·0	110·0	117·3	125·3
31·7	72·5	83·6	88·6	90·9	93·7	95·3	97·3	99·4	101·0	102·8	103·3
27·1	72·5	82·3	89·7	93·7	96·7	98·9	101·5	103·5	105·5	110·0	118·1
23·3	69·7	81·9	87·5	91·8	96·1	99·6	101·4	103·9	106·9	110·0	117·0
Im Mittel	68·0	78·8	86·6	91·1	94·7	97·8	100·4	103·0	105·6	109·2	114·0

Die Höhenprozente stimmen hier zwar bei den verschiedenen Beständen nicht mehr so gut überein wie bei den Durchmesserprozenten, allein es tritt die Gesetzmäßigkeit auch hier klar zu Tage. Konstruieren wir die Kurve, die sich aus diesen Zahlen ergibt, indem wir wieder die Stammzahlenprozente als Abszissen, die Höhenprozente, respektive die Höhenreduktionszahl R_h als Ordinaten betrachten, so erhalten wir die Fig. 3.

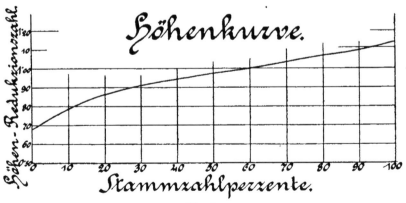

Fig. 3.

Diese Kurve hat offenbar denselben Bau wie die Durchmesserkurve, nur ist ihr Verlauf flacher. Die zugehörige allgemeine Gleichung ist:

$$R_h = 0.68 + 0.0134\,N_p - 0.000206\,N_p^2 + 0.0000012\,N_p^3 \dots 2.$$

Die mit dieser Formel berechneten Höhenreduktionszahlen sind

Stammzahlenprozente

0	10	20	30	40	50	60	70	80	90	100

Reduktionszahlen bezogen auf die Bestandesmittelhöhe

0·68	0·794	0·875	0·929	0·962	0·985	1·002	1·021	1·055	1·092	1·160

Auch hier genügt die Formel, um die aus den Bestandesaufnahmen gefundenen Mittelwerte mit annähernder Sicherheit zu beschreiben. Bemerkenswert ist, daß auch die mittlere Bestandeshöhe nahe bei 60 Prozent der Stammzahl liegt, daß also der Stamm, welcher die mittlere Grundfläche besitzt, zugleich auch Höhenmittelstamm ist. Mit Hilfe der Formel 2 ist man demnach im stande, für ein beliebiges Stammzahlenprozent die zugehörige Höhe zu berechnen, wenn nebst der Stammzahl auch die Bestandesmittelhöhe bekannt ist.

3. Die Formzahlenkurve.

Wenn wir nun der Frage „Wie gestalten sich die Formzahlen als Funktion der Stammzahlenprozente?" näher treten, so gilt es zunächst darüber schlüssig zu werden, welche der Formzahlen (Baum=, Schaft= oder Derbholzformzahl) zu wählen sei. Wenn wir jedoch berücksichtigen, daß die Baumformzahl eine überaus unbeständige und praktisch schwer zu verwertende Größe ist, daß die Derbholzformzahl einen sehr unregelmäßigen Verlauf nimmt, so bleibt uns nur die Schaftformzahl zur Wahl übrig. Da weiters aus der Schaftformzahl die praktisch wichtige Derbholzformzahl leicht zu bestimmen ist[1] und die Schaftformzahl bei gegebener Höhe und bekanntem Durchmesser ohne weiteres als Mittelwert mit Hilfe von gerade für die Fichte vorhandenen guten Formzahlentafeln — kontrolliert und korrigiert werden kann, entscheiden wir uns für die Schaftformzahl.

In der Bestandestabelle auf S. 6 sind die Bestandesschaftformzahlen nach dem Ergebnisse der Bestandesmassenaufnahme eingestellt. Diese Formzahl wurde mit jener verglichen, die sich aus Dr. Baurs Formzahlen für die Fichte Deutschlands ergibt und hieraus eine Reduktionszahl berechnet, mit welcher die Ansätze der Baurschen Tafel zu korrigieren sind. Hierbei stellte sich jedoch heraus, daß die Formzahlen Baurs auf Stammzahlenprozente angewendet bei niedrigen Prozenten zu geringe, bei sehr hohen Prozenten zu große Formzahlen geben. Ich berücksichtigte diese Fehler nach Tunlichkeit im Anhalte an zuverlässig ermittelten Stammklassenformzahlen und im Anhalte an meine Tafeln. Das Ergebnis der Formzahlenpotenzreihe als Funktion der Stammzahlenprozente ist die Kurve Fig. 4.

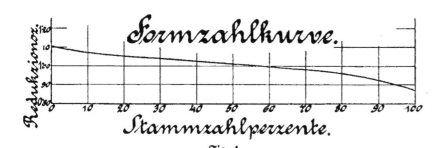

Fig. 4.

---

[1] Mitteilungen aus dem forstlichen Versuchswesen Österreichs: A. Schiffel, „Form und Inhalt der Fichte" XXIV. Heft 1899. S. 92.

Auch diese Kurve hat den gleichen Typus wie die Durchmesser= und Höhenkurve und unterscheidet sich normal nur dadurch von ihnen, daß die Orbinatenzunahme in umgekehrter Richtung erfolgt.

Die Gleichung, welche dieser Kurve annähernd entspricht, ist:

$$R_f = 0{\cdot}87 + 0{\cdot}00582 \ (100 - N_p) - 0{\cdot}0000824 \ (100 - N_p)^2 + 0{\cdot}000000477 \ (100 - N_p)^3 \ \ldots \ 3.$$

Mit dieser Formel berechnen sich die folgenden, zur Reduktion der Formzahl des Mittelstammes auf die Formzahl des bei einem bestimmten Stammzahlenprozente liegenden Stammes geeigneten Zahlen.

Stammzahlenprozente

0	10	20	30	40	50	60	70	80	90	100

Reduktionszahlen, bezogen auf die Mittelstammformzahl

1·105	1·074	1·052	1·037	1·025	1·010	1·00	0·980	0·960	0·925	0·87

Wir haben somit die Faktoren der Bestandesmassen: Durchmesser, Höhe und Formzahl als Funktionen der Stammzahlenprozente dargestellt und dabei gefunden, daß diese Faktoren die Funktionen von gleichartigen Potenzkurven sind, die durch die Gleichungen 1, 2 und 3 bestimmt sind. Wir sind demnach nunmehr in der Lage, die Masse eines beliebigen, durch das Stammzahlenprozent charakterisierten Stammes eines normalen Fichtenbestandes zu ermitteln.

4. Die Formhöhenkurve.

Wir haben jeden der drei Massenfaktoren für sich behandelt und benötigen die Formhöhe zur Massenbildung nicht. Allein es ist nicht uninteressant, auch das Verhalten der Formhöhen zu den Stammzahlenprozenten zu betrachten, zumal sich die bezügliche Kurve einfach durch Multiplikation der Gleichungen 2 und 3 ergibt. Wir ziehen es jedoch vor, anstatt mit der sich daraus ergebenden komplizierten Gleichung zu manipulieren, die Kurve aus den zahlenmäßigen Resultaten, welche wir aus der Multiplikation der beiden Reduktionszahlenreihen gewinnen, graphisch darzustellen. Hierbei ergibt sich

Stammzahlenprozente

0	10	20	30	40	50	60	70	80	90	100

Reduktionszahlen, bezogen auf die Bestandesformhöhe

0·751	0·853	0·917	0·963	0·986	0·994	1·002	1·00	1·013	1·010	1·01

Die bezügliche Kurve ist in der Fig. 5 dargestellt.

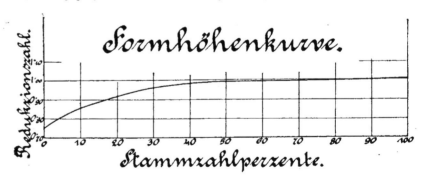

Fig. 5.

Auch diese Kurve hat noch einen, wenn auch nur schwach ausgesprochenen Wendepunkt, ist daher als Kurve 3. Ordnung zu behandeln; ihre analytische Gleichung ist annähernd:

$$R_{fh} = 0{\cdot}751 + 0{\cdot}0113\,N_p - 0{\cdot}000166\,N_p^2 + 0{\cdot}000000788\,N_p^3 \ldots 4.$$

Die mit dieser Formel berechneten Reduktionszahlen der Formhöhe sind:

S t a m m z a h l e n p r o z e n t e

0	10	20	30	40	50	60	70	80	90	100

Reduktionszahlen, bezogen auf die Bestandesformhöhe

0·751	0·848	0·917	0·962	0·987	0·999	1·002	0·999	0·996	0·998	1·01

Die Abweichungen gegenüber den aus dem Produkte h f sich ergebenden Reduktionszahlen sind sehr gering. Wie ersichtlich, ergibt sich daraus die Folgerung: Die Bestandesformhöhe steigt in normalen Fichtenbeständen vom schwächsten bis zum Mittelstamme und bleibt von dort an bis zum stärksten Stamme annähernd gleich.

5. Die Grundflächen- und Massenkurve.

Auf gleichem Wege, wie wir die Formhöhenkurve bestimmt haben, läßt sich auch die Grundflächenkurve finden. Wandeln wir zu diesem Zwecke die Durchmesserreduktionszahlen in Grundflächen um, indem wir ihr Quadrat mit $\frac{\pi}{4}$ multiplizieren, so erhalten wir die Grundflächen. Setzen wir die Grundfläche des Mittelstammes der Einheit gleich und untersuchen wir, wie viele Teile davon ein Stamm der einzelnen Stammzahlenprozente enthält, dann ergibt sich die Reduktionszahlenreihe für die Grundflächen. Diese ist:

S t a m m z a h l e n p r o z e n t e

0	10	20	30	40	50	60	70	80	90	100

G r u n d f l ä c h e n r e d u k t i o n s z a h l e n

0·308	0·475	0·595	0·702	0·802	0·913	1·02	1·17	1·37	1·64	2·43

Diese Kurve läßt sich mit einer Gleichung dritter Ordnung nicht mehr annähernd beschreiben. Verzichtet man jedoch auf die annähernde Übereinstimmung der Reduktionszahlen für die Prozente von 95 bis 100, d. i. für die stärksten Stämme, so gibt die Formel:

$$R_g = 0{\cdot}308 + 0{\cdot}0165\,N_p - 0{\cdot}000124\,N_p^2 + 0{\cdot}0000000129\,N_p^4 \ldots 5$$

annähernd die Reduktionszahlen für die Grundflächen. Dieser Ausdruck ließe sich zur rechnerischen Bestimmung der Stärkeklassen-Grundflächen-Mittelstämme aus der von der Kurve und den Achsen eingeschlossenen Fläche verwenden. Wir haben jedoch den empirischen Weg vorgezogen, weil die genaue Beschreibung der Flächenreduktionszahlen durch diese Formel nicht gewährleistet ist und der Gebrauch eines komplizierteren Ausdruckes keine Vorteile mehr bieten würde.

Multipliziert man obige Reduktionszahlen mit den Formhöhen-Reduktionszahlen, so erhält man die Massenkurve. Diese ist demnach

S t a m m z a h l e n p r o z e n t e

0	10	20	30	40	50	60	70	80	90	100

Reduktionszahlen für die Schaftmasse

0·231	0·405	0·545	0·676	0·791	0·907	1·02	1·17	1·38	1·65	2·45

Die Vergleichung der Grundflächenkurve mit der Massenkurve ergibt, daß beide Kurven vom Mittelstamme aufwärts nahezu zusammenfallen. Hieraus ergibt sich der Satz: In normalen Fichtenbeständen sind vom Mittelstamme aufwärts die Grundflächen den Schaftmassen proportional. Dieser Satz gilt jedoch nicht für die Bestandesteile, welche vor dem mittleren Durchmesser liegen. Nachstehende Fig. 18, in welcher die Massenlinie R_v voll ausgezogen ist, veranschaulicht den Verlauf der beiden Kurven graphisch.

Auch die Massen-Reduktionszahlenkurve läßt sich gleich der Grundflächenkurve erst durch eine Gleichung vierter Ordnung annähernd darstellen. Ihre Gleichung ist:

$$R_v = 0.231 + 0.0187\, N_p + 0.000141\, N_p^2 - 0.0000000134\, N_p^4 \ldots 6.$$

Die Gleichung gibt die Reduktionszahlen bis 95 Prozent der Stammzahlen annähernd richtig; in den letzten 5 Prozent sind die mit der Formel berechneten Resultate zu gering. Eine größere Annäherung könnte erst durch eine fünfgliederige Formel erreicht werden.

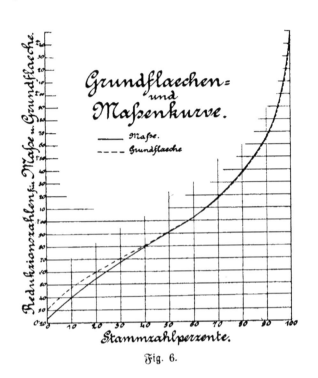

Fig. 6.

6. Folgerungen.

Aus allen diesen vorgeführten Beziehungen der Massenfaktoren zum Mittelstamme läßt sich ein gemeinsames allgemeines Gesetz formulieren. Dieses lautet: In normalen Fichtenbeständen kommt dem durch die Lage bei einem bestimmten Prozentsatze der Stammzahl gekennzeichneten Stamme ein bestimmter, in jedem Bestande gleicher Anteil von der Masse, Grundfläche, Höhe, Formzahl oder Formhöhe zu. Es verhalten sich also die Masse, Höhe, Grundfläche, Formzahl oder Formhöhe zweier Stämme eines Bestandes wie ihre Reduktionszahlen. Es sei beispielsweise die Schaftmasse des Mittelstammes v, die Schaftmasse zweier anderer Stämme v_n, beziehungsweise v_m, ihre Reduktionszahlen R_n, beziehungsweise R_m, so ist:

$$v_n = v\, R_n,\quad v_m = v\, R_m;\quad \frac{v_n}{v_m} = \frac{R_n}{R_m}.$$

Analoges gilt auch für alle Massenfaktoren.

In zwei verschiedenen Beständen verhalten sich die Massen, Formhöhen, Grundflächen, Formzahlen oder Höhen zweier Stämme gleicher Stammzahlprozente wie die Massen, Formhöhen, Grundflächen, Formzahlen oder Höhen ihrer Bestandesmittelstämme. Es sei v beziehungsweise v_1 die Masse der Mittelstämme,

v_n beziehungsweise v_n^1 die Masse der Stämme zweier Stämme gleicher Stamm
zahlenprozente in zwei verschiedenen Beständen, so ist:

$$v_n = v\,R_n,\quad v_n^1 = v_1\,R_n;\quad \frac{v_n}{v_n^1} = \frac{v}{v_1}.$$

Diese Gesetze lassen sich mit der gleichen Sicherheit nicht ableiten, wenn
man den Mittelstamm durch einen beliebigen anderen substituieren, d. h. die Re=
duktionszahlen auf einen beliebigen anderen, bei einem bestimmten Prozentsatze
der Stammzahl gelegenen Stamm beziehen wollte, weil die Beständigkeit der
Reduktionszahlen verschiedener Bestände am größten ist, wenn man sie auf den
Mittelstamm bezieht.

Es ist also der Mittelstamm, d. i. seine Masse und seine Massenfaktoren
charakteristisch für den Bestand. Die innere Struktur und Eigenart gelangt im
Mittelstamme und der Stammzahl zum Ausdruck; in diesen ist alles enthalten,
was Kopezky mit Zuwachskoeffizient und Zuwachscharakteristik bezeichnet.

Hiermit wäre die Darstellung aller Beziehungen, welche zwischen den
Massenfaktoren eines beliebigen Bestandesgliedes zu den Massenfaktoren des Be=
standesmittelstammes bestehen, erschöpft und es erübrigt uns nur noch die An=
wendung dieser Gesetze an einem Beispiele zu demonstrieren. Nehmen wir an, es sei in
einem Bestande die Stammzahl $N = 860$, die Mittelstärke $d = 25\cdot8\,cm$, die
Mittelhöhe $h = 26\cdot6\,m$, die Schaftmasse $V = 620\,m^3$ gegeben und es handle
sich darum, die Schaftmasse derart zu zerlegen, daß eine Sortierung und Be=
wertung ermöglicht werde. Ein Prozent der Stammzahl beträgt $8\cdot6$ Stämme.
Es unterläge keiner Schwierigkeit, die mittlere Höhe, Grundfläche, Formzahl und
Schaftmasse aus den vorbehandelten Kurven für jedes einzelne $8\cdot6$ Stämme um=
fassende Prozent zu berechnen, oder den graphisch dargestellten Kurven direkt zu
entnehmen; praktisch hätte jedoch eine derartige Detaillierung keinen Sinn, weil
die Sortiments= und Wertsunterschiede erst bei größeren Abstufungen eintreten.
Wir werden also Stammklassen bilden und wählen als für diesen Zweck ent=
sprechender, nicht die Urichsche, sondern die R. Hartigsche Stammklasseneinteilung.[1]
Nehmen wir eine Einteilung zu fünf Klassen mit gleicher Flächenverteilung, so liegt
nach S. 199 der Mittelstamm jeder Stärkeklasse und zwar

in der Stärkeklasse I   II   III   IV   V
bei dem Prozent der Stammzahl:

| 17 | 40 | 68 | 84 | 95 |

Die Stärkeklasse umfaßt Prozente der Stammzahl

| 35·5 | 22·8 | 17·7 | 13·5 | 10 5 |

Es entfallen sonach in jede Stärkeklasse Stämme:
$860\,p = N_n =$

| 305 | 196 | 153 | 116 | 90 |

Die Masse des Mittelstammes ist: $v = 0\cdot721\,m^3$, die Formzahl des Schaftes:
$f = 0\cdot519$. Die Reduktionszahlen bei obigen Prozenten sind:

In der Stärkeklasse I   II   III   IV   V
Reduktionszahlen der Masse R_v

| 0·51 | 0·79 | 1·13 | 1·47 | 1·83 |

Reduktionszahlen der Höhe R_h

| 0·5 | 0·94 | 1·03 | 1·07 | 1.12 |

Die Schaftmasse der Mittelstämme beträgt demnach $v\,R_v = v_n$

| 0·368 | 0·569 | 0·815 | 1·06 | 1·32 |

Die Schaftmasse der Stärkeklasse beträgt $N_n\,v_n$

| 112 | 112 | 125 | 123 | 119 m^3 |

---

[1] Vgl. „Über Bestandeshöhen und Bestandesformzahlen". Z. f. b. g. F. 1900.
S. 20 u. ff.

Die Mittelhöhe h R_h jeder Stärkeklasse beträgt

22·6 | 25·0 | 27·3 | 28·5 | 29·8 m

Reduktionszahlen R_d der Durchmesser

0·76 | 0·89 | 1·06 | 1·21 | 1·37

Der Mittendurchmesser R_d d jeder Stammklasse beträgt

19·6 | 23·0 | 27·3 | 31·2 | 35·3 cm.

Werden die Massen der einzelnen Stammklassen addiert, so soll das Resultat der Bestandesmasse gleich sein. In diesem Falle gibt die Summe 591 m³, daher um 29 m³ oder um 4·7 Prozent zu wenig. Die Kontrolle läßt sich auch mit der Grundfläche durchführen.

Wie ersichtlich, ist es in befriedigender Weise gelungen, die gegebene Schaftmasse zu zerlegen. Wir haben ganz entsprechende Abstufungen in den Durchmessern und Höhen und auch die zugehörigen Stammzahlen gewonnen; es ist also die Sortimentsbildung und deren Bewertung auf einer weit breiteren Grundlage tunlich, als dies aus den Daten des Mittelstammes allein möglich wäre. Wer sich damit noch nicht begnügen will, dem liefern die im XXIV. Hefte der „Mitteilungen aus dem forstlichen Versuchswesen Österreichs" enthaltenen Tabellen noch weitere Behelfe zur Sortimentenbildung. Mit den Daten Schaftmasse, Höhe und Durchmesser findet man nämlich in der dort enthaltenen Kubierungstabelle auch die Durchmesser in $\frac{h}{4}$, $\frac{h}{2}$ und $\frac{3}{4}$ h und in den Tabellen 3, 4 und 5 auch alle Daten, um den Schaft auf Derbholzlänge zu kürzen und die Schaftmasse in Derbholzmasse umzuwandeln.

Bei der praktischen Verwertung der hier auf Grund einer Reihe zweckmäßig ausgewählter Bestände dargestellten Gesetze darf nicht außer acht gelassen werden, daß diese Gesetze die Beziehungen der Massenfaktoren zum Mittelstamme und seinen Massenkomponenten nur annähernd beschreiben und von der Wirklichkeit umsomehr abweichen werden, je mehr sich die Bestandesbeschaffenheit von der normalen entfernt.

In der Übereinstimmung der Massen oder Grundflächen, die sich aus der Zerlegung ergeben, mit der zu zerlegenden Grundfläche oder Masse des Bestandes liegt jedenfalls eine erwünschte Kontrolle darüber, ob sich der Bestand den Gesetzen fügt. Ist man in der Lage, eine genügende Anzahl von Probestämmen zu fällen, so wird es wohl niemanden einfallen, den Bestand in die Zwangsjacke irgend welcher Gesetze zu pressen, sondern man wird im Bestande selbst alle jene Daten erheben, welche der Zweck, dem die Bestandesaufnahme dienen soll, erheischt; man wird auf diese Weise der Individualität des Bestandes Rechnung tragen und kann sich alle möglichen Kurven, die benötigt werden, konstruieren, d. h. man kann die Gesetze des Bestandes als Spezialfall darstellen.

Ich fasse daher die praktische Verwertung dieser Gesetze nur für solche Fälle ins Auge, wo Zeit und Umstände eine exakte Bestandesaufnahme nicht zulassen. In diesen Fällen werden diese Gesetze einen Behelf dazu bieten, um die eingehendere Bestandesaufnahme zu ersetzen.

Ich habe schon angedeutet, daß die praktische Verwertung vornehmlich in zwei Kategorien von Fällen tunlich erscheint.

1. Bei allen Bestandesaufnahmen, die sich in kurzer, gegebener Zeit vollziehen müssen. Bei Gutsverkäufen, Nachlaßverhandlungen, bei Expertisen zum Zwecke des Abschlusses von Holzabstockungsverträgen oder Verkäufen nach der Fläche. In allen diesen Fällen wird man die Arbeiten im Walde rasch in dem nur unbedingt erforderlichen Ausmaße vollziehen müssen. Da es zur Anwendung des Verfahrens genügt, wenn Stammzahl, Durchmesser, Höhe und Formzahl des Mittelstammes bekannt sind, so ist das Minimum der Anforderungen in der Auskluppierung einer Probe, Bestimmung des Kreisflächenmittelstammes und in dem Messen einiger Höhen von solchen Stämmen gegeben, die den Mittelstamm-

durchmesser besitzen. Aus den gemessenen Höhen wird das Mittel gebildet und mit dem Eingange Höhe und Durchmesser die Formzahl aus einer Formzahlen=tafel bestimmt. Es ist also nicht erforderlich, einen Probestamm zu fällen und dennoch sind alle Daten gegeben, um bei Anwendung des beschriebenen Ver=fahrens eine eingehende Bestandesaufnahme zu ersetzen.

2. Bei allen Wertsberechnungen, die auf Grund von gegebenen Ertrags=tafeln erfolgen müssen. Es sind dies: Waldwertsberechnungen, Umtriebszeitbestim=mungen, forststatische Untersuchungen, auch Weiserprozent= oder Wertzuwachsprozent=berechnungen. Da alle modernen Ertragstafeln die zur Anwendung des Verfah=rens der Massenzerlegung erforderlichen Daten enthalten, reduziert sich die Arbeit auf das im vorgeführten Beispiele gezeigte Ausmaß.

Voraussichtlich gelten diese, aus normalen Fichtenbeständen abgeleiteten Ge=setze allgemein, das ist für alle Holzarten, wenn auch die Kurven selbst, oder die Kurvenformeln verschiedene sein mögen. Ich betrachte es als einen Vorzug des hier beschriebenen Verfahrens, daß es allen Holzarten, insbesondere auch allen durch lokal gebräuchliche Methoden in der Begründung und Erziehung der Be=stände hervorgerufenen Eigentümlichkeiten in der Bestandesbeschaffenheit angepaßt werden kann. Zur Bestimmung des Verlaufes der gesuchten Kurven genügen wenige, die lokalen Eigentümlichkeiten charakteristisch zum Ausdruck bringende Bestände. Es ist selbstverständlich nicht erforderlich, für die gefundenen Kurven einen mathematischen Ausdruck zu suchen, weil das ganze Verfahren auch lediglich graphisch angewendet werden kann.

K. u. k. Hofbuchdruckerei Carl Fromme in Wien.

CPSIA information can be obtained
at www.ICGtesting.com
Printed in the USA
LVIC06n0212141117
556209LV00008B/78

Basic Penmanship

Basic Penmanship

John Lancaster

Dryad Press Ltd, London

This book is dedicated to Horace Firth a well-loved and respected gentleman from Yorkshire and a 'real' father

I am most grateful to friends and others who have helped me to compile this book. These include John Milton-Smith, Lilly Lee, Jean Larcher, Anne Hechle and John Poole, calligraphers and illuminators, and John Davie for his excellent photography. My appreciation also goes to William F Waller, my editor, for his encouragement and help

John Lancaster.

Cheltenham
January 1987

ISBN 0 8521 9677 6

Typeset by
Servis Filmsetting Ltd, Manchester
and printed in Great Britain by
The Bath Press Ltd, Bath, Avon
for the publishers
Dryad Press Ltd
8 Cavendish Square
London W1M 0AJ

Contents

Introduction

Penmanship, often referred to as 'calligraphy', is the result of an acquired craftsmanship with pens and inks and writing surfaces. It is really an aspect of drawing in which undeviating mark-making becomes a form of artistic expression. In order to produce beautiful letters, therefore, the calligrapher's gestures must be as near to perfection as possible.

Craftsmanship of high quality takes time to achieve; it demands endless practice and results from what eventually appears to be a natural skill, and that of the calligrapher is no exception. The penman must know his tools and materials, and how he might handle these in combination, both to transmit written ideas and to encase these with a visual beauty which will give pleasure to the reader.

The practice of penmanship is a discipline which is best appreciated through personal experience. It provides untold satisfaction in the making of meaningful gestures while leading to a greater understanding and delight in the work of fine letterers and scribes of earlier generations.

1. The Scribe's Basic Equipment & Materials

Table or desk
Table-top easel
Writing board
Pens
A pencil
A ruler

Inks
A brush
Water jar
Pen jar
Papers and cards
Container for writing materials

This short list contains only the essential materials. Once the calligrapher has gained some experience, more sophisticated equipment will be required. Let us look briefly at each item.

Table or desk

A table or desk on which to write is essential—make sure it does not wobble.

Instead of buying a table improvise by placing a sheet of blockboard or other suitable material over a couple of supports (trestles, small filing cabinets or wooden boxes).

Table-top easel

This can be of great help to the beginner in that it gives him or her a strong support. Some calligraphers prefer to write on a board which rests on their laps. (DART and Stable-mate are reliable brands.)

Writing board

The writing board is one of the calligrapher's most vital pieces of equipment.

It should be of a reasonable size to suit your own requirements (ie 24in × 18in [60cm × 45cm] is convenient for general use). Again you can improvise and make your own from sheets of blockboard, hardboard, formica offcuts or even strong card.

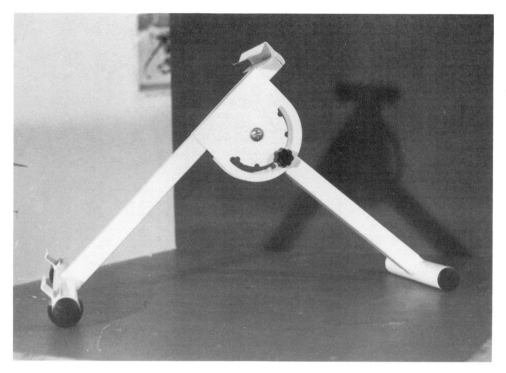

Fig. 1 *This photograph of the DART table-easel, which is made by Pentographics Ltd (see list of suppliers), clearly shows the mechanism by which the angle of the writing board may be adjusted. The drawing board (not included here) slips under the clip (top) which is, in fact, on the adjustable extending arm, and is pushed upwards until the board can be rested upon the two clips (bottom left).*

I find my own DART table easel strong, firm and clearly well made and finished.

Pens

These are obviously the most essential items and I suggest a range consisting of those listed below.

Fibre-tipped pens (Edding 1255 Calligraphy Pens, black and coloured), (Berol 051 Italic Pens: sizes 1.5mm, 2.0mm, 3.5mm and 5.0mm) and (Rexel Calligraphy Pens Nos 3.5 or 5).

Steel nibs (William Mitchell Pens, Nos 1–3).

Pen holders for the above (usually referred to as 'dip pens') together with small brass reservoirs (which fit beneath the nibs and act as small ink containers).

Calligraphic fountain pens (Winsor and Newton, Margros and Platignum).

A pencil

Used for lining up sheets of paper or card prior to beginning work; pencils should be hard (H or 2H) rather than soft (HB or B).

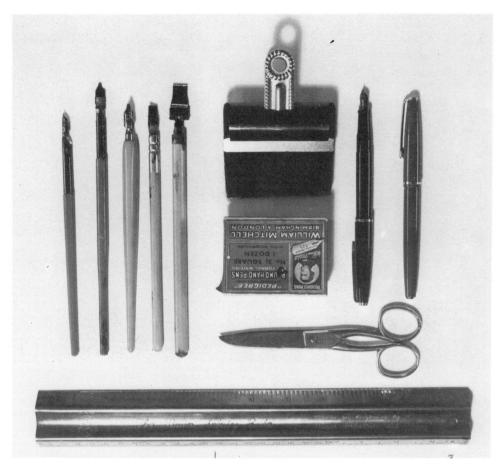

Fig. 2 *Basic equipment used by the calligrapher.*

A ruler

Another useful piece of equipment, the ruler, may be of wood, metal or plastic material, and of adequate length (I like a 'Veteran' plastic ruler which is 18in [45cm] long).

Inks

Inks for use with steel nibs, should be restricted to good quality fountain-pen ink (Quink, or similar, black and/or coloured) which is ideal for practice work.

For permanent pieces of writing ink should be black, *non-waterproof* Chinese Stick Ink (hard sticks of ink which, when rubbed in a little water, will produce a free-flowing black ink) (liquid form in bottles from Winsor and Newton, or 'Youth' Chinese Ink CP 305).

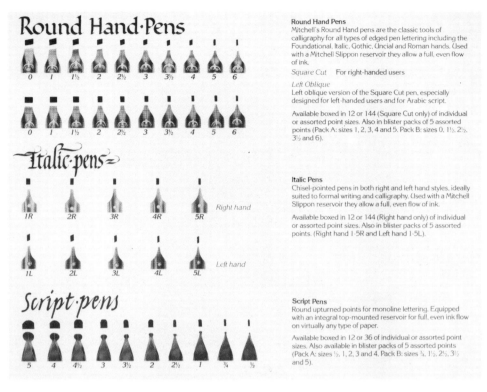

Round Hand Pens

Mitchell's Round Hand pens are the classic tools of calligraphy for all types of edged pen lettering including the Foundational, Italic, Gothic, Uncial and Roman hands. Used with a Mitchell Slippon reservoir they allow a full, even flow of ink.

Square Cut For right-handed users

Left Oblique

Left oblique version of the Square Cut pen, especially designed for left-handed users and for Arabic script.

Available boxed in 12 or 144 (Square Cut only) of individual or assorted point sizes. Also in blister packs of 5 assorted points (Pack A: sizes 1, 2, 3, 4 and 5. Pack B: sizes 0, 1½, 2½, 3½ and 6).

Italic Pens

Chisel-pointed pens in both right and left hand styles, ideally suited to formal writing and calligraphy. Used with a Mitchell Slippon reservoir they allow a full, even flow of ink.

Available boxed in 12 or 144 (Right hand only) of individual or assorted point sizes. Also in blister packs of 5 assorted points. (Right hand 1-5R and Left hand 1-5L).

Script Pens

Round upturned points for monoline lettering. Equipped with an integral top-mounted reservoir for full, even ink flow on virtually any type of paper.

Available boxed in 12 or 36 of individual or assorted point sizes. Also available in blister packs of 5 assorted points (Pack A: sizes ½, 1, 2, 3 and 4. Pack B: sizes ¾, 1½, 2½, 3½ and 5).

Fig. 3 *Pens (Courtesy Rexel).*

A brush

A brush is necessary for replenishing your dip pen (ie used to place ink into the brass reservoir beneath the nib). Use a cheap type (sable), No 4, 5 or 6.

Water jar

It is vital to keep a water jar (or water container of some sort) by your side at all times when working for cleaning pens and brushes.

Pen jar

A similar receptacle in which to store pens upright can be very helpful; it will enable you to keep your pens together so that you can locate the one required next without difficulty.

Papers and cards

These are the 'supports' on which the calligrapher often writes and, as a beginner, you will certainly use paper.

For practice this paper can simply be typing paper but for more developed work should be hand-made (preferably hot-pressed).

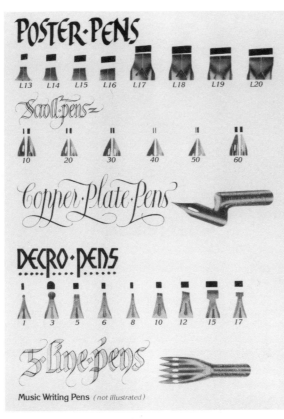

POSTER·PENS

L13 L14 L15 L16 L17 L18 L19 L20

Scroll·Pens

10 20 30 40 50 60

Copper·Plate·Pens

DECRO·PENS

1 3 5 6 8 10 12 15 17

5 line·pens

Music Writing Pens *(not illustrated)*

Poster Pens
Large broad point edge pens for poster and display lettering. Slightly canted to the left with upturned edge to facilitate lettering on textured surfaces. Each pen is equipped with an integral reservoir.

Available boxed in 12 individual or assorted point sizes. Also available in blister packs of 5 assorted points (Sizes L13, L14, L15, L17 and L19).

Scroll Pens
Double edged pen for interesting double line calligraphic effects and decorative borders. Cannot be used with a reservoir.

Available boxed in 12 individual or assorted point sizes. Also available in blister packs of 5 assorted points (Sizes 10, 20, 40, 50 and 60).

Copperplate Pens
Very fine, responsive elbow oblique pens for Copperplate writing. Cannot be used with a reservoir. Available in one size only.

Available boxed in 12's or in blister packs of 5.

Decro Pens
Generally preferred by more advanced calligraphers to produce a variety of decorative lettering styles. Each pen is equipped with an integral top mounted reservoir.

Available boxed in 12 individual or assorted point sizes. Also available in blister packs of 5 assorted points (Sizes 1, 3, 5, 12 and 17).

Five Line & Music Writing Pens
Calligraphers use these pens to achieve interesting graphic effects. The Five Line pen may also be used for scraperboard and for ruling musical staves. The 0268 Music Writing pen is a recent reintroduction to the range and is a very versatile calligraphic tool.

Available only on blister cards containing 2 Five Line and 3 Music Writing pens.

Joseph Gillott Drawing and Mapping Pens
The specialised series of artists pens under the Gillott brand have, for many years, been recognised as the world's leading drawing pens. They are the automatic choice for

Joseph Gillott

1290 170 1950 290 291 303 404 1068A

659 850 2788

pen and ink sketching, mapping, engraving, extra fine writing and general drawing. Each is made from the finest Sheffield steel, hand-finished and carefully tested for top quality performance.

Drawing Pens
Available boxed in 36 or 144 of an individual point size. Also available in blister packs of 5 assorted points (Sizes 170, 290, 303, 404 and 1950).

Mapping Pens
Available boxed in 36 or 144 of an individual point size. Also available in blister packs of 5 assorted points – sizes 659(2), 850(1) and 2788(2).

Slippon Reservoirs
Slippon reservoirs need to be attached beneath the nib, the tip just touching the point, about ⅛" from the end. Ensure that the tongue of the reservoir is making light contact with the pen *(See illustration)*.

Pen Holders
Three pen holders are available:
Multipurpose, for use with all the William Mitchell pens with the exception of the Gillott drawing and mapping pens.
Mapping, for use with the Gillott Crow Quill pens (659, 850 and 2788).
Drawing, for use with the Gillott drawing pens (190, 290, 291, 303, 404, 1068A, 1290 and 1950).

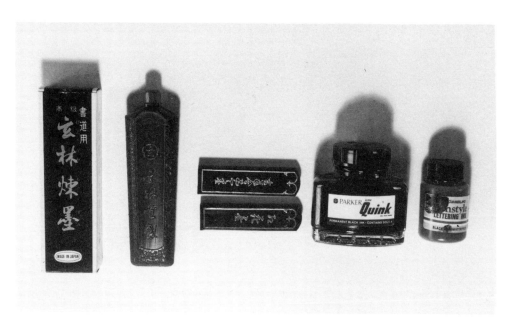

Fig. 4 *Inks.*

Fig. 5 *Here I am writing with a goose quill on a board which rests on my lap. The photograph was taken by John Davie in my studio at home. I am sitting so that the light from the window comes from the left so that no shadows are cast across my work surface.*

Fig. 6 *This photograph of Anne Hechle, the well-known British calligrapher who works in Bath, shows her range of materials and equipment—a table-easel, an abundance of quills (possibly goose and swan feathers), an Anglepoise-type lamp which supplies her with working light onto her board, paints, palettes, brushes and jars of raising preparations, gums and pigments.*

If you employ card (for instance when producing notices) then you should select these from the excellent ranges in local art materials shops.

Container for writing materials

A helpful piece of equipment in which to store pens, pen-holders, pencil, brush, etc; it can be improvised from a cardboard box, wooden cigar box, typing paper box or a simple plastic box such as an ice-cream container.

Additions to this basic kit might be:
  rubber (eraser); pair of scissors; masking tape; glue; stanley knife (or similar); drawing or mapping pins; ball-point pens; an Angle-poise lamp (or similar).

Advanced work is done on vellum in the form of calfskin, sheepskin and goatskin. This requires knowledge of how to prepare and handle it and should not be attempted at this stage.

2. The Writing Position

It is important to establish a good writing position. This should be one in which the calligrapher is comfortable and relaxed, and which will enable him or her to work for long periods of time. For this reason it is unwise to write on an horizontal table-top.

I suggest that you try *two* positions which are admirable for calligraphy—both of which I employ depending on the type of work I am doing.

1. Sit facing your desk or table. Rest your board on your lap and let it rest against your desk or table at an angle of approximately 45 degrees to the horizontal. (Obviously this angle can vary to suit your own requirements.)

Tuck the right arm into your side and allow your wrist and hand to rest gently upon the sloping work-surface (the board). This will give you the best and most comfortable position at which to write.

2. Sit facing your table-easel, which should be placed on your desk or table and, if necessary, be anchored to it with a small 'G'-clamp so that it will not move inadvertently as you write.

Slope the easel at an angle of some 30–40 degrees for a sensible and comfortable position.

Rest your right hand on the sloping work-surface at the most comfortable position *for you*, remembering that this might be retained for a considerable period of time.

As an aid to good calligraphic practice I suggest that you place a sheet of strong cartridge paper or white card across your writing board so that it covers the surface of the board from the position where you intend to write downwards. This may be pinned or taped to the edges of the board and will provide a 'practice area' on which to test out pens. It will also protect the work you are doing as this will slip behind it easily. It is wise to see that this is at right-angles to the left- and right-hand edges of your lettering board.

Fig. 7 *The basic writing position.*

Figs. 8 and 9 *The basic writing position shown from different angles.*

Fig. 9

Figs. 10 and 11 *The basic writing position when using a table-top easel.*

Fig. 11

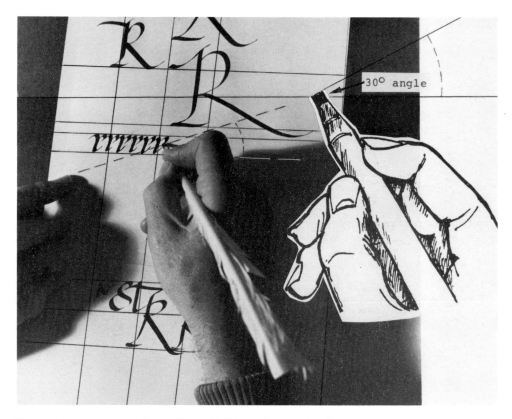

30° angle

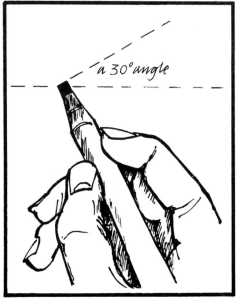

a 30° angle

Fig. 12 *The character of the basic round-hand script advocated in this book—whether written with a fibre-tipped pen or, as in this photograph, with a swan quill—is achieved by holding the pen at a constant angle of 30 degrees to the horizontal writing line. This follows traditional methods adopted by scribes throughout many centuries.*

Fig. 13

Fig. 14 *This rather interesting two-line effect resulted accidentally from a dryish fibre-tipped pen.*

3. Formal Script Letters

It can be somewhat difficult for a beginner to control a dip pen satisfactorily and I therefore recommend the use of a fibre-tipped pen instead. The one employed in writing the letters in this section was a Rexel Calligraphy Pen No 5.0. It is an easy pen to use and I find that the fibre point is firm so that both strong and quite fine lines result.

Pen-made letters are not mechanical and stilted. The fact that they are made by human beings using simple points and ink gives them a lively, individualistic quality so that they appear 'to dance' or 'to sing' to their

Fig. 15 *Detail of a piece of formal script from a notice written by the author.*

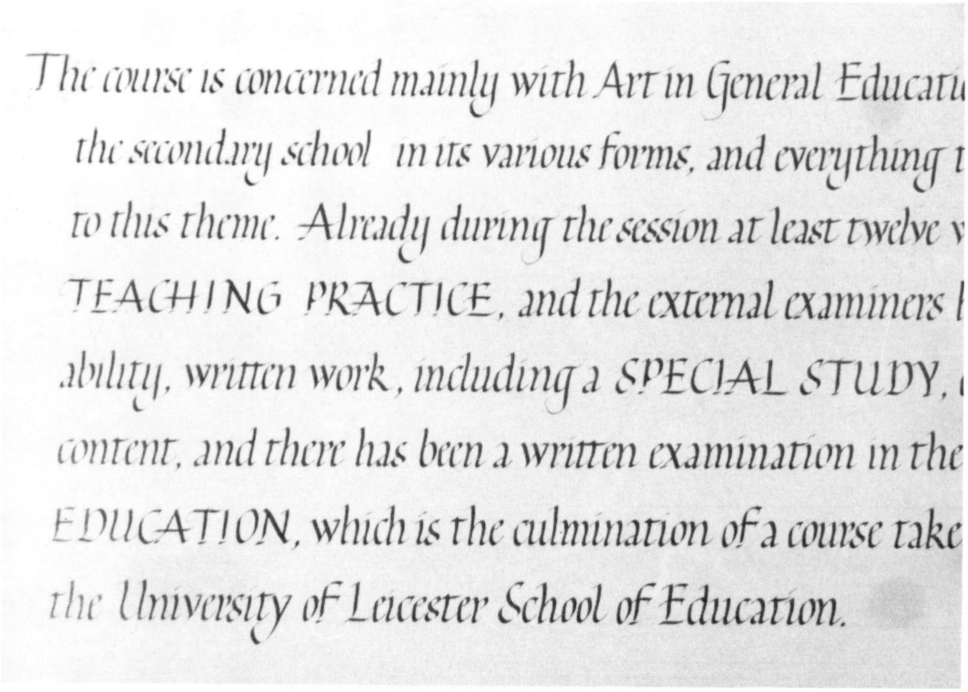

The course is concerned mainly with Art in General Educatu
the secondary school in its various forms, and everything t
to this theme. Already during the session at least twelve v
TEACHING PRACTICE, and the external examiners
ability, written work, including a SPECIAL STUDY,
content, and there has been a written examination in the
EDUCATION, which is the culmination of a course take
the University of Leicester School of Education.

readers with what can only be described as a spontaneous beauty. I have produced the following illustrations with the help of guide-lines in order to show each letter's structure more clearly, and do so with the help of a circle, an idea explained in my book *Calligraphy Techniques*, pp 43–50. (See Useful Books, p. 89).

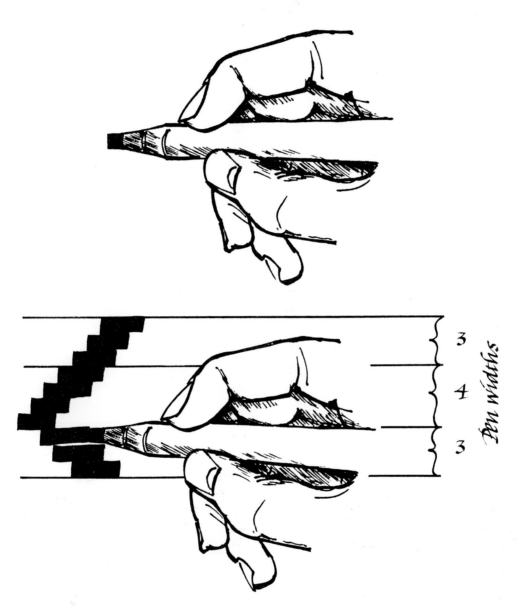

Fig. 16 *To determine the size of your letters turn your pen sideways, as shown here, and employ 'pen widths' as basic units. Various combinations of such units will allow you to alter the proportions and, in consequence, the characters of individual letters.*

Fig. 17 *These examples of lower-case letters, are based (from left to right) upon four, five, six and seven pen widths (basic units).*

Fig. 18 *In this example the selected capital letters are based (from left to right) upon* four, five, six and seven pen widths.

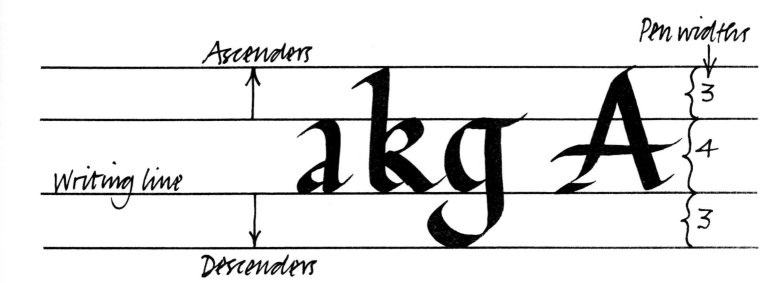

Ascenders

Pen widths

Writing line

Descenders

3
4
3

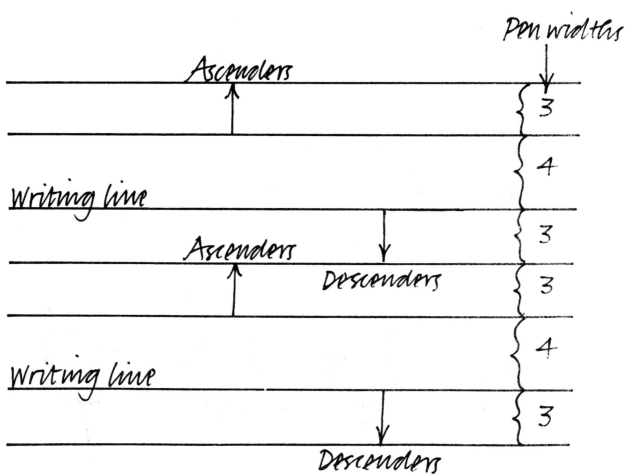

Pen widths

Ascenders

Writing line

Ascenders

Descenders

Writing line

Descenders

3
4
3
3
4
3

Fig. 19 *Here I have tried to show how a few selected letters relate to the writing line, with ascenders going up to the height of the capital letters. In a page of lettering the ascenders and descenders hit a common line.*

Fig. 20 *(Opposite)*

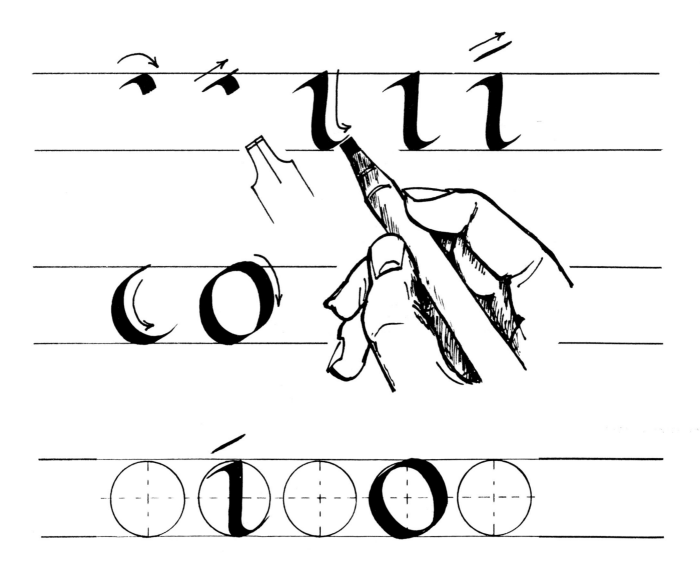

Fig. 21 *The letters of the alphabet are made up of straight lines and curves, and at first it is suggested that you practise the lower-case letters 'i' and 'o', which are given on this page. These two forms characterise the majority of letters.*

Use a fibre-tipped pen (Rexel Calligraphy Pen—either a No. 3.5 or a No. 5). Dip pens and quills are for more advanced work.

Retain a constant pen angle as this will determine the characters of the letters.

Follow the arrows.

30

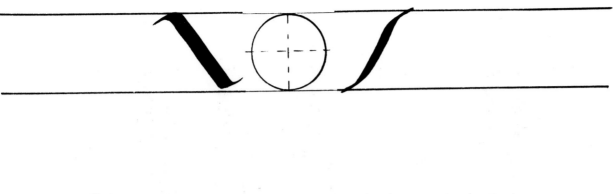

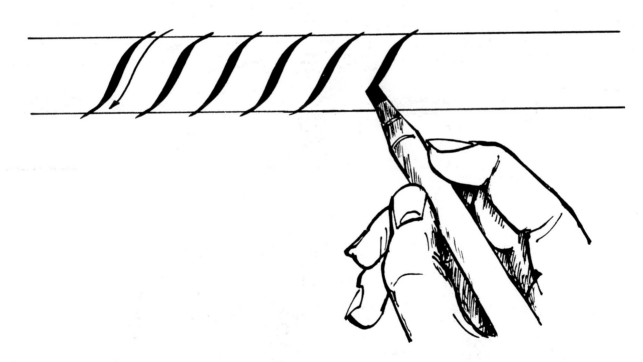

Fig. 22 *Lines which slope from left to right or from right to left are also important and need to be practised separately as shown here. A circle can be a helpful guide for most letters when you are beginning to learn how to do script.*

Fig. 23

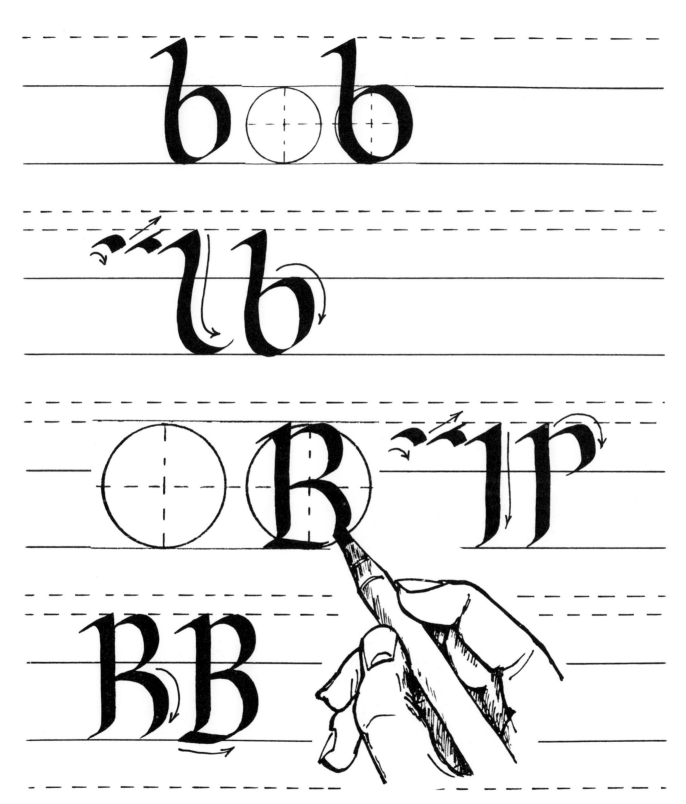

Fig. 24

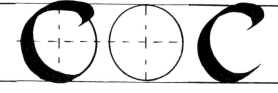

A 'round' letter that is helped by the circle guide.

Fig. 25

34

Fig. 26

A round lower-case 'e' is contrasted, in this case, with an angular 'Classical' capital letter.

Fig. 27

36

Fig. 28

Fig. 29

38

Notice the strong 'round' arch.

Fig. 30

39

Fig. 31

40

Fig. 32

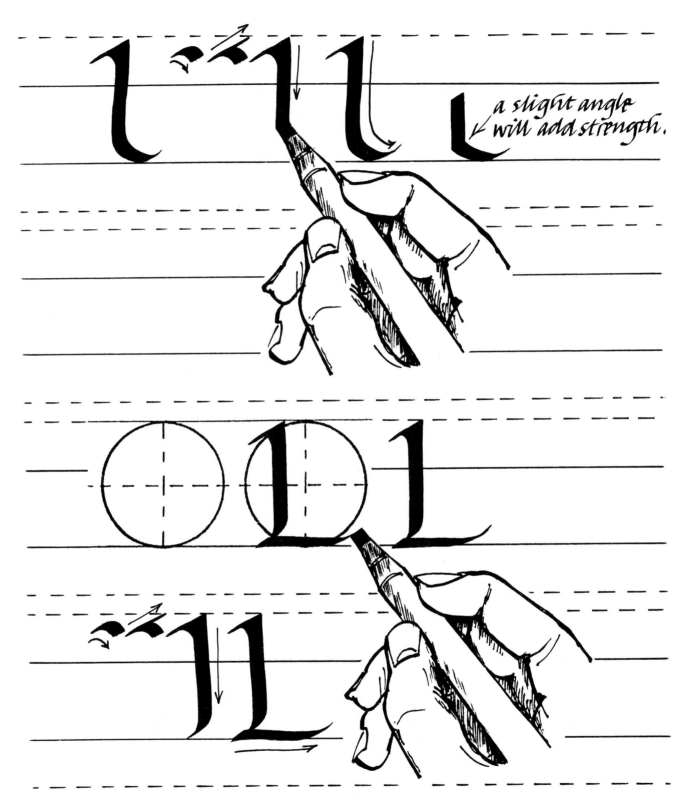

a slight angle will add strength.

Fig. 33

42

This should be a strong 'Roman' arch.

and not a weak hook

Fig. 34

43

Fig. 35

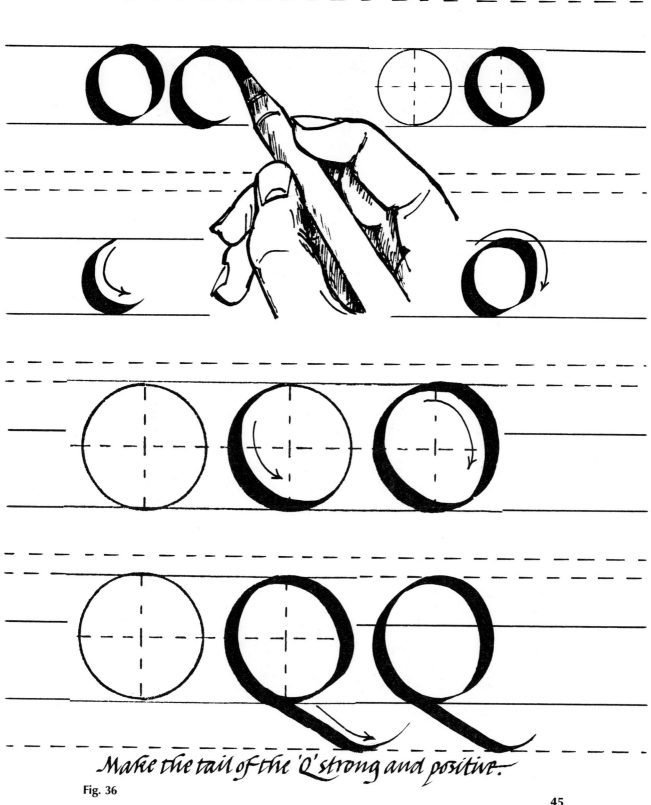

Make the tail of the 'Q' strong and positive.

Fig. 36

Fig. 37

Fig. 38

Fig. 40

Fig. 41

A strong, reversed 'Roman' arch is best

not an awkward hook.

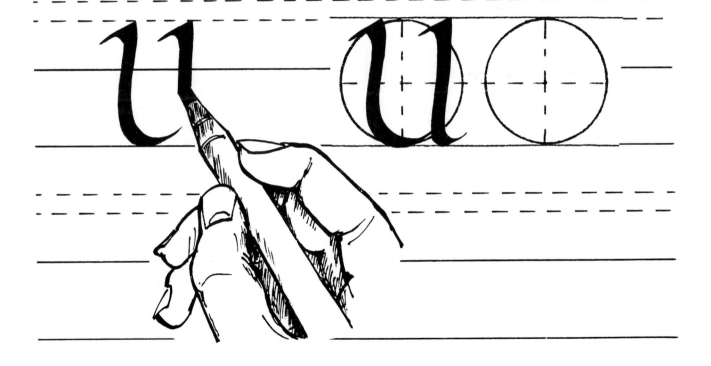

Fig. 42

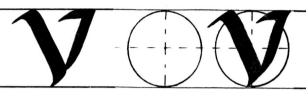

Starting directly above the bottom of the first stroke.

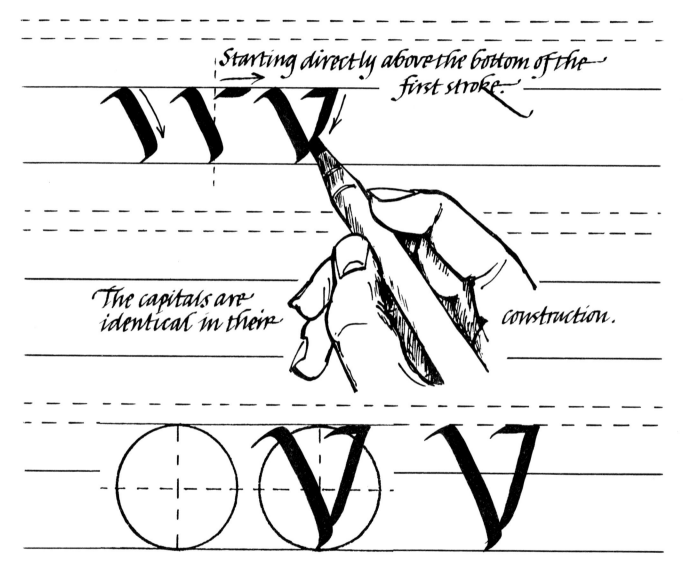

The capitals are identical in their construction.

Fig. 43

52

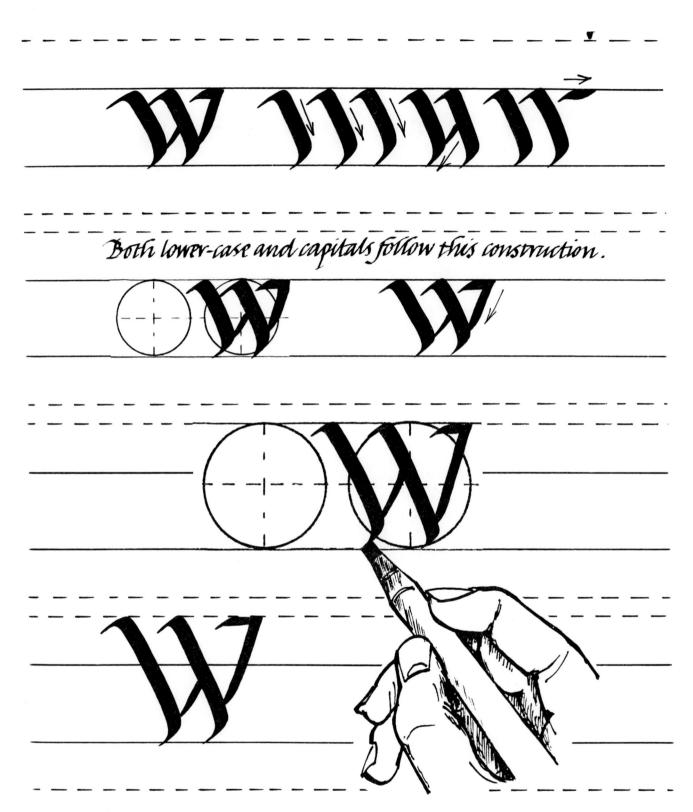

Both lower-case and capitals follow this construction.

Fig. 44

These capitals
are simply larger
versions of
lower-case letters.

Fig. 45

54

Fig. 46

55

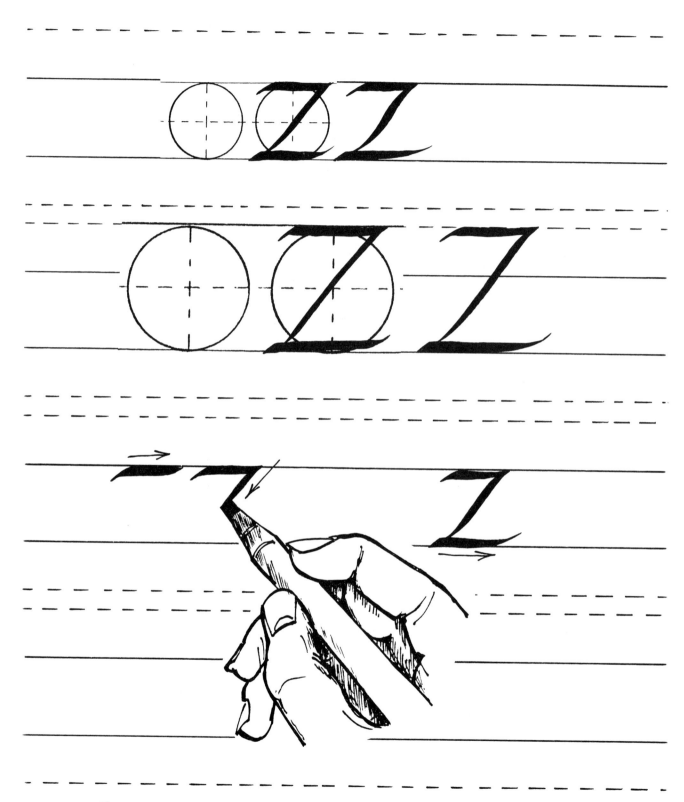

Fig. 47

abcdefghijklmnop
qrstuvwxyz
ABCDEFGHIJK
LMNOPQRSTUV
WXYZ &&

Fig. 48 *Lower-case and capital round-hand alphabets written quickly without using guidelines.*

4. Some Historically-based Scripts

This book is concerned primarily with 'formal' writing and therefore its main emphasis is on a *basic round-hand* script and how the beginner might learn the rudiments of shaping letters with a square-ended pen; an implement that has been employed over many centuries by calligraphers.

At this point you might find it interesting to look at some of the scripts which were used during the Medieval period—principally between the seventh and fifteenth centuries, and a few from the twentieth century. Although various styles tend to be clearly represented it must be recognised that a gradual metamorphosis took place as individual letters changed and their characters evolved.

The historical examples which are used here—unless captioned otherwise—have been written freely with a Rexel fibred-tipped pen (No 5.0) held at a constant angle to the horizontal writing line. Please note that once determined the writing angle never changed. As alternatives to the Rexel pen I could have used a *quill*, or a *reed*, a William and Mitchell metal nib, an Edding 1255 Calligraphy Pen or a Berol Italic Pen.

Writing was normally contained in books which, unlike paintings and sculptures, were relatively easy to transport from place-to-place. Monks or other religious persons would take them along when they travelled around between monasteries, abbeys and cathedrals in this and other countries. The scribes would look at such books with great interest and different lettering styles would be examined or possibly copied freely. This meant that regional styles exerted a gradual influence upon those in other places, adding a unique richness to the calligraphy produced at a particular period in history.

Throughout the Middle Ages manuscript pages were alive with pattern, gold, colour and masses of hand-written text. The scribes and illuminators— the *artists* and *craftsmen* who produced them—worked in the monastic scriptoria, and these artistically-talented people were allotted individual work-spaces. These would contain a desk as well as writing and illuminating materials so that they could get on with the task of making prayer books, religious treatise, bibles, bestiaries, letters and legal documents. Many of these works of art were, in fact, simply everyday artifacts.

abcdesqrlx

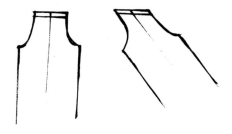

49 *'Half-uncial' letters of the seventh and eighth centuries* AD *were produced with a quill or reed pen which was cut obliquely. They were characterised by thick vertical and thin horitzonal strokes.*

INCOXSRAG

Fig. 50 *'Uncial' capital letters which are typical of the sixth- and seventh-century* Celtic manuscripts. They are full and round Majascule letters.

pharesautemgc
nuitesrom
csromautemge
nuitaram
aramautemcenuit
aminadab
aminadabautem
cenuitnasson
naassonautemge
nuitsalmon

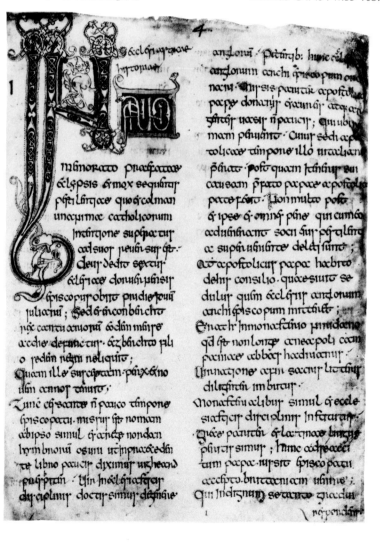

60

Fig. 53 *Insular 'half-uncials' written by Eadfrith in the late seventh century with an Anglo-Saxon translation inserted in the tenth century. Source: the Lindisfarne Gospels—British Museum MS Nero D 1V.*

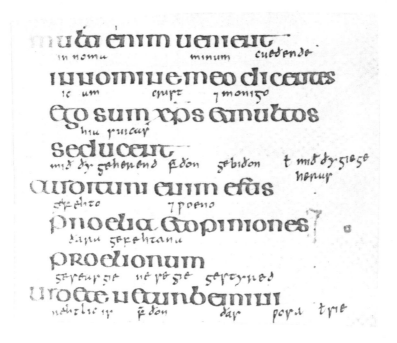

Fig. 54 *Lower-case letters based on tenth-century book hands. They are all round in shape, strong and firm and are characterised by an open beauty of form.*

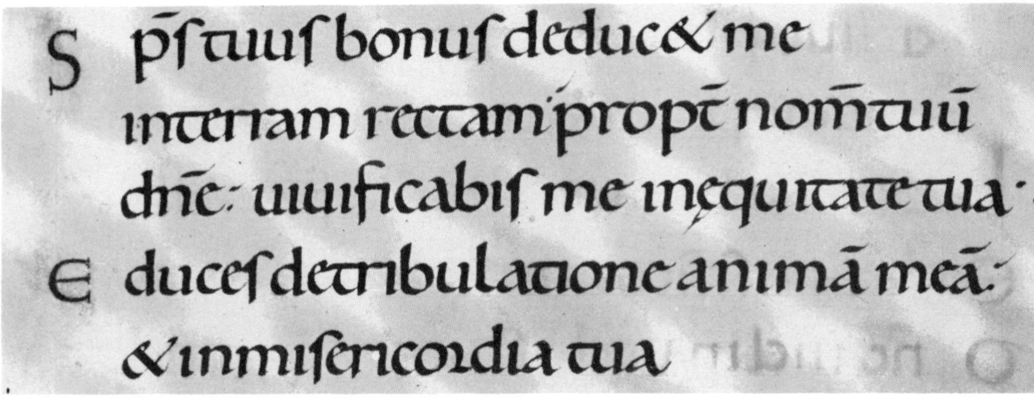

ſ pſ tuuſ bonuſ deduc& me
intterram rectam propt nomtuiu
dñc: uiuificabiſ me inęquitate tua·.,
∈ duceſ detribulatione animā meā:
&inmiſericoidia tua

Fig. 55 *Pen lettering of the tenth century from the scriptorium at Winchester. Courtesy of the British Museum Harl. MS 2904.*

abcderqxτmp

Fig. 56 *Examples of letters belonging to the late tenth and early eleventh centuries.*

abcdeglmopstu

Fig. 57 *These letters from a twelfth-century manuscript show that they are becoming slightly angular while losing their roundness.*

62

Fig. 58 *Writing from an early thirteenth-century manuscript. Source: Royal MS 3C V111 f2— British Library BL/C/MS/ 115.*

63

abcdefghmorst

Fig. 59 *'Gothic' black letters dating from the thirteenth and fourteenth centuries.*

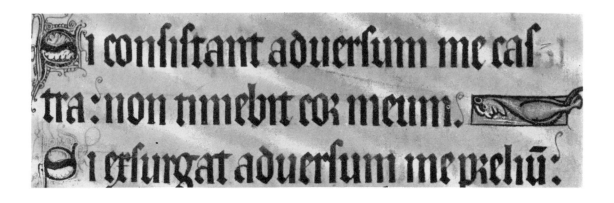

Fig. 60 *Fourteenth-century 'Gothic' or 'Black' lettering from the Luttrel Psalter (British Museum).*

abcdefghijkxyz

Fig. 61 *Modern (twentieth-century) formal round-hand based upon tenth-century scripts.*

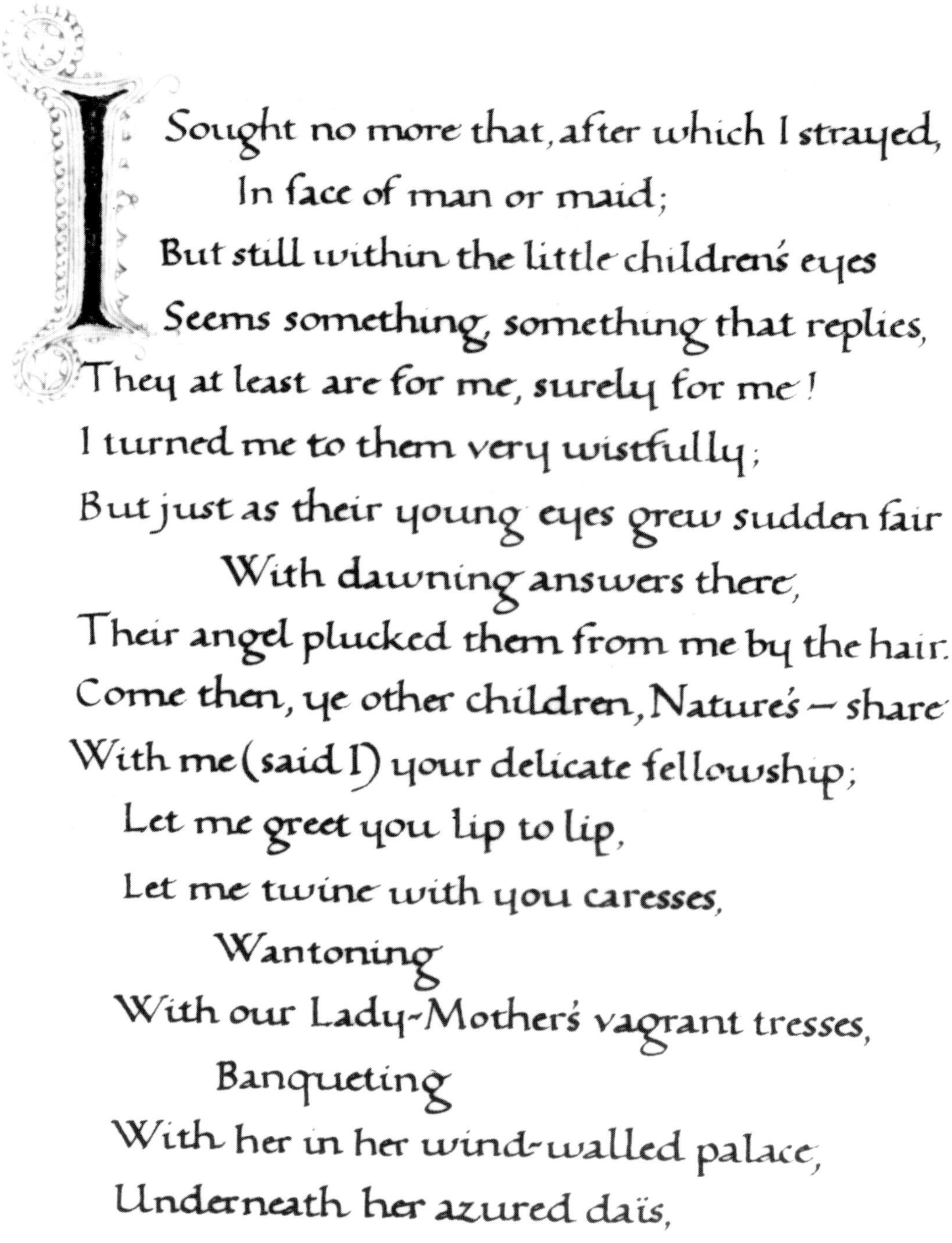

I Sought no more that, after which I strayed,
In face of man or maid;
But still within the little children's eyes
Seems something, something that replies,
They at least are for me, surely for me!
I turned me to them very wistfully;
But just as their young eyes grew sudden fair
With dawning answers there,
Their angel plucked them from me by the hair.
Come then, ye other children, Nature's — share
With me (said I) your delicate fellowship;
Let me greet you lip to lip,
Let me twine with you caresses,
Wantoning
With our Lady-Mother's vagrant tresses,
Banqueting
With her in her wind-walled palace,
Underneath her azured daïs,

Fig. 62 *Page from* The Hound of Heaven *by Francis Thompson, written in a modern round hand by Graily Hewitt in 1906.*

Courtesy of the Victoria and Albert Museum L91 L 326-1958.

66

Fig. 63 *The writing on this envelope,
although an italic hand, shows evidence of its
development from Medieval writing.*

5. Design & Layout

In the Middle Ages the design and layout of manuscript books followed a basic, pragmatic pattern. This meant that the outside (left- and right-hand) margins had to be wide enough to both facilitate the handling of areas of written text without smudging, and to allow lettering and illuminations to be set off in comfort.

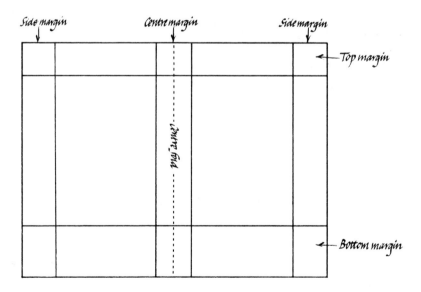

Fig. 64 *The margins of a page-opening should follow a simple pattern, ie* top margin *smaller than* bottom margin *(possibly half), and the* side margins *equal to the* centre margin. *Note that the* bottom margin *is larger than a side margin. It should also be* pointed out that the fold in the centre of the page-opening bisects the *centre margin, and that the* side margins *should be wide enough to facilitate easy handling of the book.*

Fig. 65 *This example shows the layout of one page from a French illuminated manuscript of the early fifteenth century. The Gothic script is enveloped by a richly decorated border based upon ivy-leaf decoration. Courtesy of the British Museum Addl. MS 16997 F63 B99 (Hours of E Chevalier).*

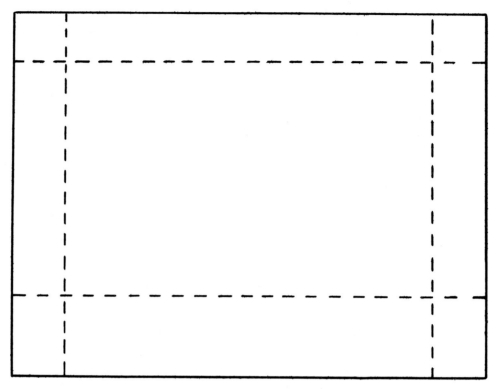

Fig. 66 *A panel of lettering—known as a broadside—can be set out, as in this example, with the top and side margins* *approximately the same but with the bottom margin larger.*

Figures 64 and 66 and their accompanying captions show the relative spacings of margins to areas of text, and should you go on to design page-openings or even a book then they might act as a useful guide.

Fig. 67 *This 'vertical' example of a broadside demonstrates the use of formal scripts and decorative features. Reproduced by gracious permission of H M The Queen. Ref SO4119—Lord Chamberlain, St James's Palace, SW1).*

THE LOYAL & DVTIFVL ADDRESS OF THE LONDON COVNTY COVNCIL

TO THE QUEEN'S MOST EXCELLENT MAJESTY

May it please your Majesty

WE, the Chairman, Aldermen and Councillors of the London County Council — representatives of the people of your Majesty's capital — humbly approach your Majesty to express our profound sorrow at the passing of his late Majesty King George the Sixth, and respectfully to extend to you our loyal congratulations on your Accession to the Throne.

Our late and beloved King was to us an example of all that is good in public and in family life. His unfailing kindliness and ready sympathy, his courage and resolution throughout the anxious and troubled years of his reign and the warm encouragement which at all times he gave to us in our work for the people of his capital, endeared him to us and inspired a deep sense of gratitude and admiration.

The people of London have lost a true friend. They derive great comfort, however, from their certain knowledge that his noble spirit and his beneficent influence endure and today live in the person of your Majesty.

It is with gratitude and sincere appreciation that we recall your Majesty's interest in our activities and your visits to our County Hall in the years before your Accession. We have particularly happy memories of your Majesty's gracious presence and the presence of your Royal husband, the Duke of Edinburgh, at the celebration of the Council's Diamond Jubilee and at the opening of our Royal Festival Hall.

We are proud to stand before your Majesty on this day and to offer with our most loyal congratulations on your Majesty's Accession to the Throne our warm affection and our homage.

We pray that your Majesty, His Royal Highness and your family may be blessed with long life and every happiness and that the years of your Majesty's reign may be marked by the increase of freedom, peace and prosperity throughout the world.

Edwin Bayliss
Chairman of the Council.

Countersigned
Clerk of the Council.

22nd May 1952

Left margin (scrolls):
LONDON thou art of townes A per se
RICHEST
Soveraign of cities seemliest in sight
IN
Of high renoun riches & royaltie
BONTIE
Of lordis barons & many a gudly knyght
AND
EMPRESS of townes exalt in honour
IN
In beawtie berying the crone imperiall
BEWTIE
SWEET paradise precelling in pleasure
CLEAR
London thou art the flour of Cities all
William Dunbar

Right margin (scrolls):
Strong be thy wallis that about the standis
AND
Wise be the people that within the dwellis
EVERIE
Fresh is thy ryver with his lusty strandis
VERTEW
Blith be thy chirches wele sownyng be thy bellis
THAT
Rich be thy merchauntis in substaunce that excellis
IS
Fair be their wives right lovesome white & small
WENIT
Clere be thy virgyns lusty under kellis
DEAR
London thou art the flour of Cities all
&W. Scribe

SWEET ROIS OF VERTEW & OF GENTILNES

DELYTSVM LILIV OF EVERIE LVSTYNES

6. The Evolution of Letters

We can go back as far as 20,000 years BC in the development of human consciousness to see that, at that stage, *picture writing* was beginning to be an important means of visual communication in which 'pictures' of objects, animals and human beings were used as a means of transmitting messages and stories. The letters of the alphabet, as we know them today, gradually evolved over many centuries from such picture writing through, broadly speaking, *Egyptian* (approximately 3,000 BC); *Phoenician, Greek, Etruscan* and *Roman* (approximately 1,000 BC—AD 200); *Medieval* (approximately AD 400–1,400) and on to what I shall refer to as *Modern* (twentieth century.) A much more detailed account is given in my book *Lettering Techniques* (see Useful Books, p. 89), in the form of an extended table, which might be of interest to the reader.

Here I have provided some examples of modern script letters while attempting to show how these have evolved. I hope that this will give the reader an overall idea while sharpening his or her interest in a fascinating subject.

I suggest that if you wish to make a more detailed study then it might be worthwhile to pursue the subject further by reading some more specialised books. Those by Fairbank (1970 and 1975) and Jackson (1981), which are given in Useful Books, p. 89 are excellent.

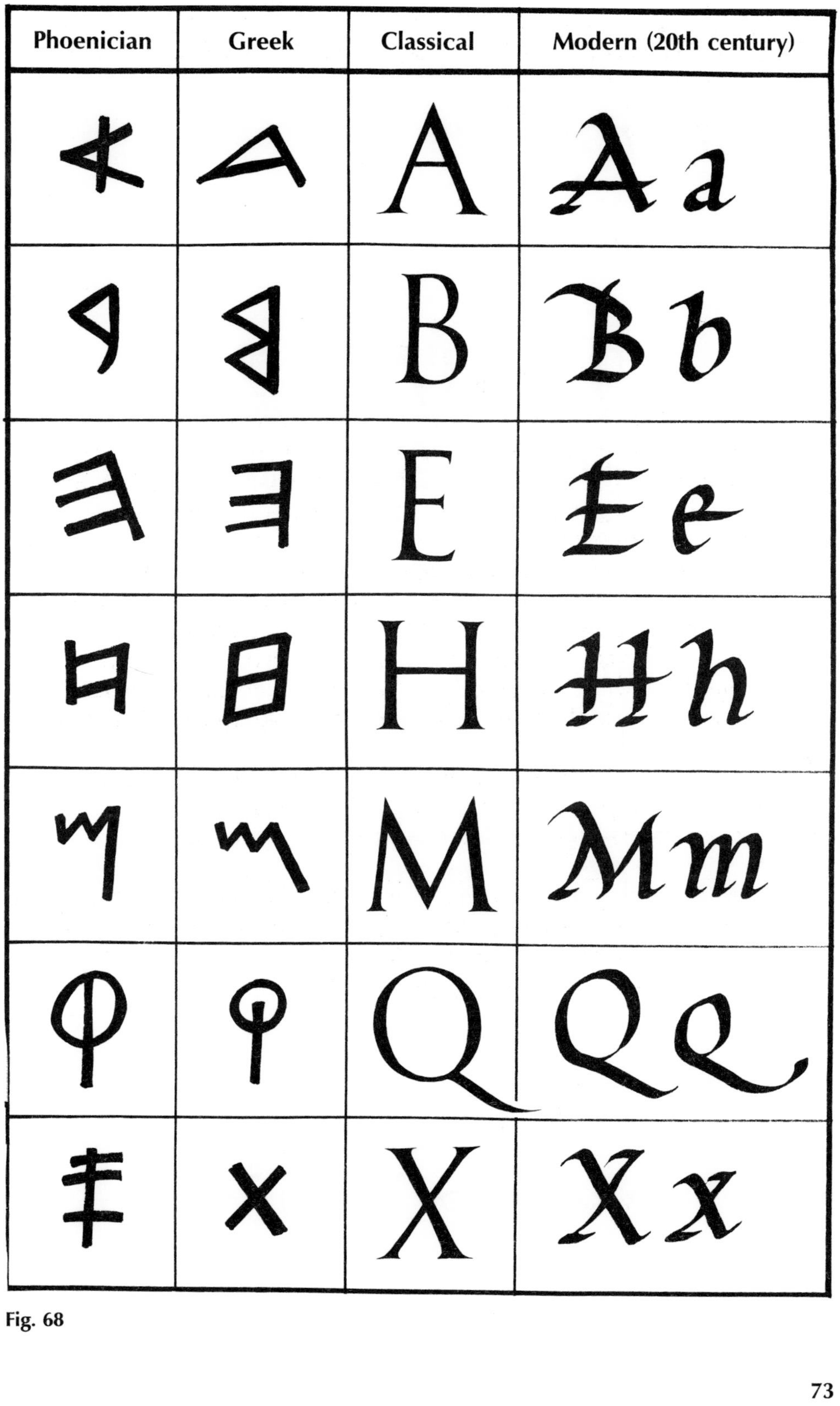

Phoenician	Greek	Classical	Modern (20th century)
		A	A a
		B	B b
		E	E e
		H	H h
		M	M m
		Q	Q q
		X	X x

Fig. 68

7. Some Examples of Calligraphy

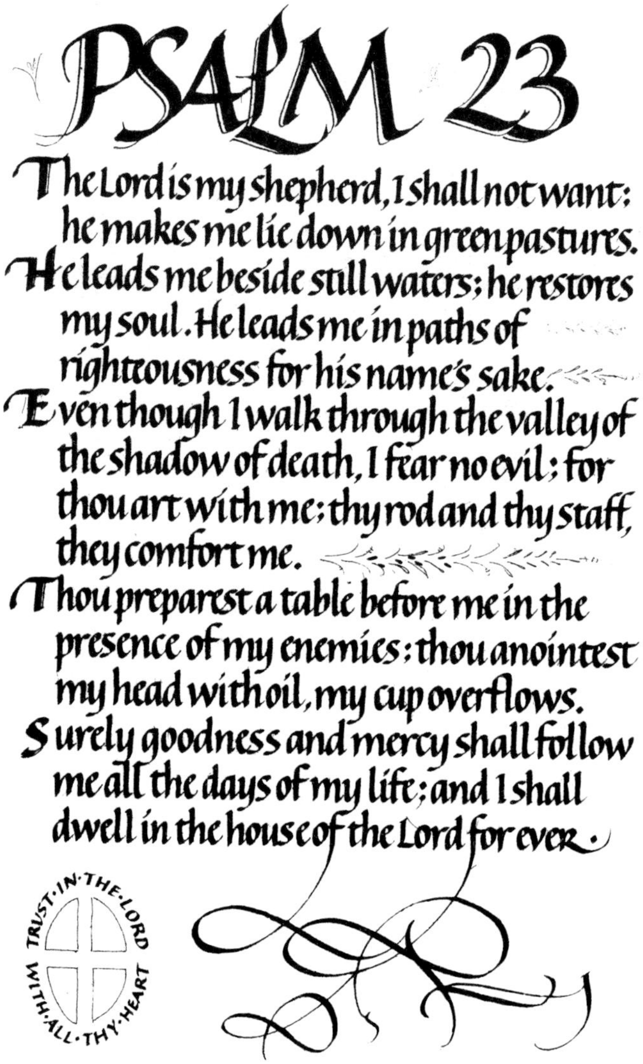

PSALM 23

The Lord is my shepherd, I shall not want:
he makes me lie down in green pastures.
He leads me beside still waters; he restores
my soul. He leads me in paths of
righteousness for his name's sake.
Even though I walk through the valley of
the shadow of death, I fear no evil; for
thou art with me; thy rod and thy staff,
they comfort me.
Thou preparest a table before me in the
presence of my enemies; thou anointest
my head with oil, my cup overflows.
Surely goodness and mercy shall follow
me all the days of my life; and I shall
dwell in the house of the Lord for ever.

TRUST·IN·THE·LORD·WITH·ALL·THY·HEART

Fig. 69 *Calligraphy by Lilly Lee.*

74

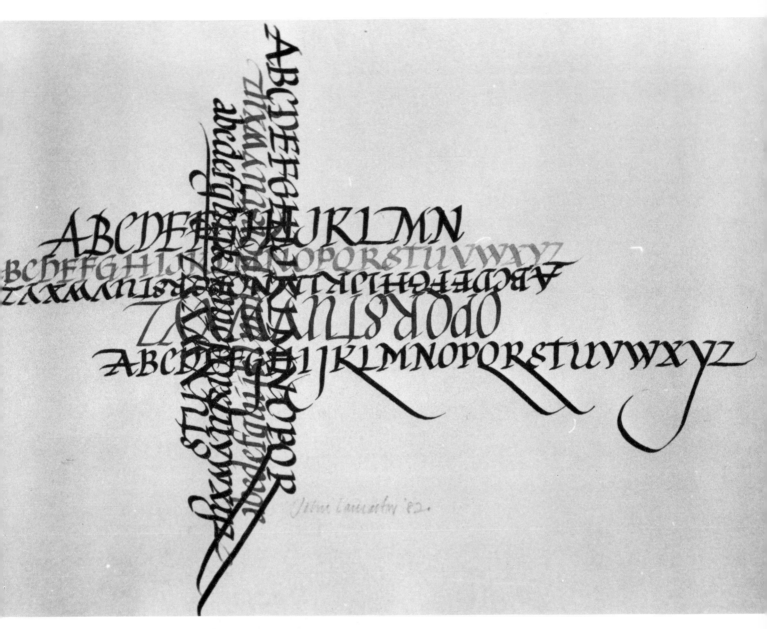

Fig. 70 *Freely-written alphabets—John Lancaster.*

75

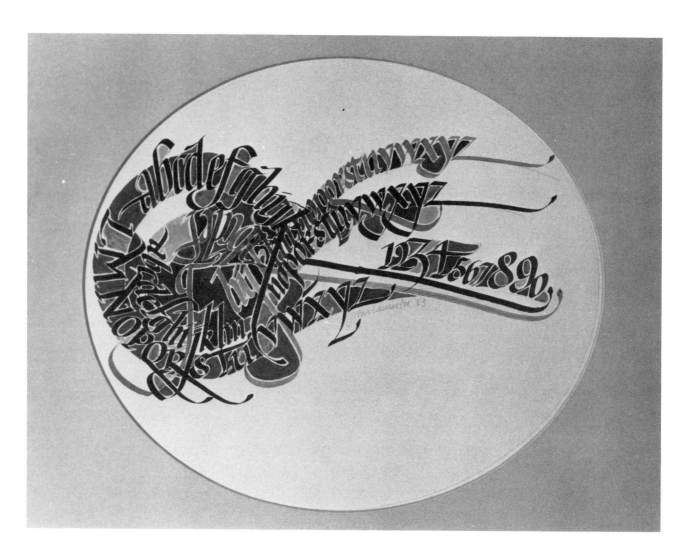

Fig. 71 *Decorated panel in script and illuminating colours by John Lancaster.*

CITY OF LEEDS

AT A MEETING OF THE COUNCIL OF THE CITY OF LEEDS held in the Council Chamber at the Civic Hall on Wednesday, the Fourth day of February in the year Nineteen hundred and Eighty-one, as a Special Council convened for the purpose, THE LORD MAYOR [Mr. Councillor Eric Atkinson, M.B.E.] in the chair,

It was Resolved unanimously:-

"THAT under and in pursuance of the powers conferred by Section 249 [5] of the Local Government Act 1972 the Council admit

HENRY SPENCER MOORE O.M., C.H.

to be an

HONORARY FREEMAN OF THE CITY OF LEEDS in recognition of the long and distinguished services rendered by him to the arts and his benefaction to the City."

Dated this Third Day of July, Nineteen hundred and Eighty-one.

Lord Mayor

Chief Officer and Director of Administration.

E. J. Milton Smith.

CONFERMENT OF THE ROLE OF HONORARY FREEDOM OF THE CITY OF LEEDS ON HENRY SPENCER MOORE O.M., C.H. 3RD JULY 1981

Fig. 72 *An illuminated address in ink, raised gold and illuminator's colours on vellum by John Milton-Smith. Tooled leather case by Joan Webster.*

Fig. 73 *Pen lettering by Jean Larcher.*

Christchurch Craft Society

EXHIBITION

EXHIBITION HALL TOWN HALL IPSWICH

Wednesday-Saturday 3rd-13th November

Open 9am-5pm everyday except Fridays when the exhibition closes at 8pm and Saturdays at 4pm

Fig. 74 *An exhibition notice by John Poole.*

Fig. 75 *Double pen script written by John Poole.*

Fig. 76 *A poster designed and written by John Poole.*

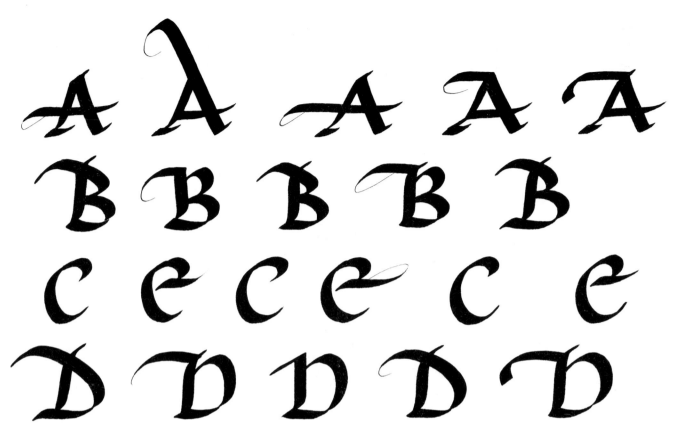

Fig. 77 *In this—and the next five illustrations—letters of the alphabet are shown in slightly varied forms.*

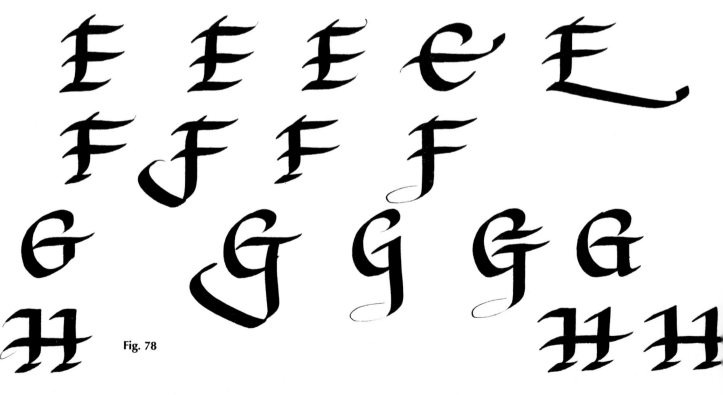

Fig. 78

82

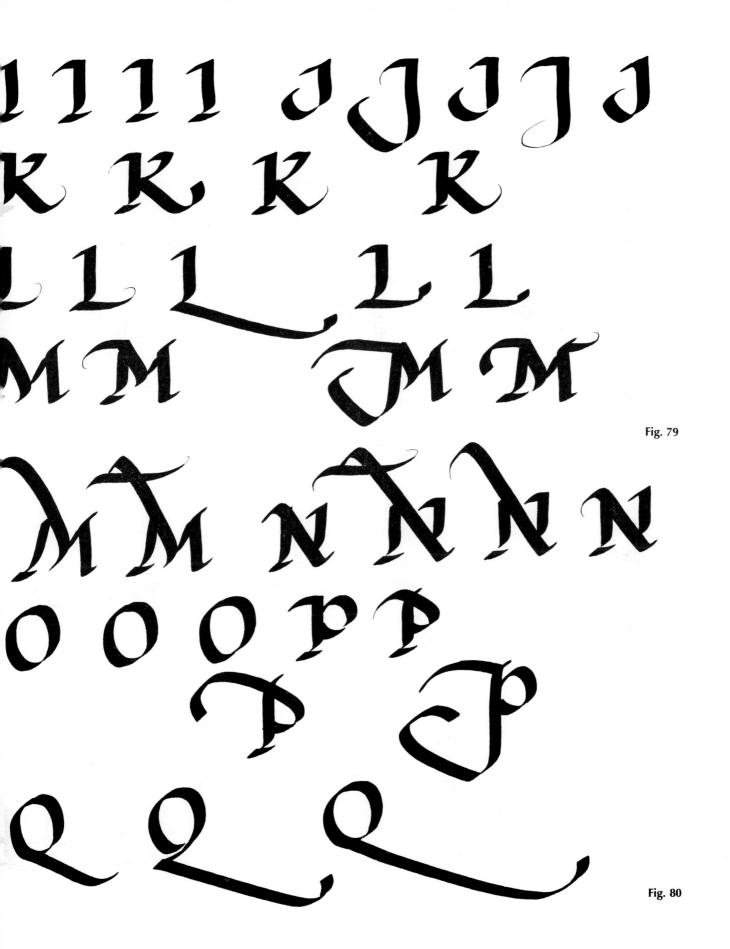

Fig. 79

Fig. 80

Fig. 81

Fig. 82

Fig. 83 *Quickly scripted letters by the author.*

Fig. 84 *A selection of flourished letters by the author.*

Fig. 85

Fig. 86 *Flourished tails.*

Useful books

Backhouse, Janet, *The Illuminated Manuscript*, Oxford, Phaidon, 1979

Baker, Arthur, *The Calligraphy Manual*, London, Dryad Press, 1984

Douglass, Ralph, *Calligraphic Lettering with Wide Pen and Brush*, London, Pitman, 1975 and New York, Watson Guptill

Fairbank, Alfred, *A Book of Scripts*, London, Penguin, 1949

Fairbank, Alfred, *A Handwriting Manual*, London, Faber, 1975

Fairbank, Alfred, *The Story of Handwriting: Origins and Development*, London, Faber, 1970

Furber, Alan, *Layout and Design for Calligraphers*, London, Dryad Press, 1984

Gourdie, Tom, *Calligraphic Styles*, London, Studio Vista, 1979

Gourdie, Tom, *The Puffin Book of Lettering*, London, Penguin, 1970

Gourdie, Tom, *The Simple Modern Hand*, London, Collins, 1975

Holme, C G (ed), *Lettering of Today*, London, Studio, 1949 and New York, Studio

Jackson, Donald, *The Story of Writing*, London, Studio Vista, 1981

Johnston, Edward, *Formal Penmanship and other papers*, London, Lund Humphries, 1980

Lamb, C M (ed), *The Calligrapher's Handbook*, London, Faber, 1956

Lancaster, John, *Lettering Techniques*, London, Batsford, 1982

Lancaster, John, *Calligraphy Techniques*, London, Batsford, 1986

Martin, Judy, *The Complete Guide to Calligraphy Techniques and Materials*, London, Phaidon, 1984

Mahoney, Dorothy, *The Craft of Calligraphy*, London, Pelham Books, 1982

Sassoon, Rosemary, *The Practical Guide to Calligraphy*, London, Thames and Hudson, 1984

Switkin, Abraham, *Hand Lettering*, New York, Harper Row, 1976

Whalley, Joyce Irene, *The Student's Guide to Western Calligraphy: An Illustrated Survey*, Shambhala, Boulder and London, 1984

Wong, Frederick, *The Complete Calligrapher*, New York, Watson-Guptill, 1980

Suppliers

You will probably find most of the equipment and materials which you require in local art materials shops and they may stock those supplied by the following companies.

Berol Ltd
Old Meadow Road
King's Lynn
Norfolk
PE30 4JR
(Papers and fibre-tipped pens)

Dryad Ltd
P.O. Box 38
Northgates
Leicester
LE1 9BU
(Calligraphy pens and inks)

Faber-Castell
D-8504, Stein/Nürnburg
W. Germany
(Broad and fine felt-tipped pens filled with paint lacquer for use on paper, wood, glass, ceramic or metal)

Falkiner Fine Papers Limited
117b Long Acre
Covent Garden
London
WC2E 9PA
(All manner of calligraphic supplies including lettering papers and books)

Margros
182 Drury Lane
London WC2
(Various writing materials)

Osmiroid
Gosport
Hampshire
(Calligraphic materials including pens, inks & papers)

Pentographics Ltd
Marsh Road
Lords Meadow Industrial Estate
Crediton
Devon
EX17 IEU
(Dart table-top easel and drawing board)

Platignum Pen Co
Mentmore Manufacturing
Platignum House
Six Hills Way
Stevenage
Herts
(Fountain pens, fibre-tipped and felt-tipped pens and inks)

Rexel Ltd
Gatehouse Road
Aylesbury
Bucks
HP19 3DT
(William Mitchell steel nibs, calligraphy pens and reservoirs)

Winsor & Newton Ltd
51 Rathbone Place
London W1
(All kinds of papers, cards, pens, inks, drawing (writing) boards and easels)

90

Index

92

MATHEMATICS ACCOMPLISHED

Y6 BOOSTER
TEACHING ASSISTANT'S BOOK

Peter Clarke

Acknowledgements

Our thanks to Jackie Trudgeon and Sheila Viegas, Year 6 teachers from Tower Hamlets, London, who provided the inspiration for this book.

Thanks also to the BEAM Development Group:
Jo Barratt, Rotherfield Primary School, Islington
Mark Day, Hanover Primary School, Islington
Catherine Horton, St Jude and St Paul's Primary School, Islington
Simone de Juan, Prior Weston Primary School, Islington
Helen Wood, Vittoria Primary School, Islington

Published by BEAM Education
Maze Workshops
72a Southgate Road
London N1 3JT
Telephone 020 7684 3323
Fax 020 7684 3334
Email info@beam.co.uk
www.beam.co.uk
© BEAM Education 2007, a division of Nelson Thornes

ISBN 978 1 906224 24 0
British Library Cataloguing-in-Publication Data
Data available
Edited by Marion Dill
Cover design by Malena Wilson-Max
Layout by Suzan Aral, Reena Kataria and Matt Carr
Printed in Spain

Contents

Introduction

HOW TO USE THIS BOOK

The *Year 6 Booster* Teacher's book contains 30 lessons for a teacher to use with a group of children. This can be the whole class or a selected group. Each lesson focuses on a specific skill or idea that children often find difficult.

Each of the 30 teacher's lessons has a follow-up activity to go with it. These follow-up activities are detailed in this book, and they are intended for you to use with a selected group of children who need more practice.

The sessions are all presented on a double page like this

Vocabulary

This is a list of key words and phrases that children need to learn, understand and use when talking about their work. Try to use these words yourself when you talk with children, in a context that shows what they mean – then the words will become familiar and meaningful to children.

Activity

This is what you actually do with the children. It may be a game, an activity or working through an activity sheet.

We have included examples of questions or prompts for you to present to children. This will help challenge their mathematical thinking further and encourage children to discuss mathematical processes.

About the maths

In this section, you will find information about the way children learn maths as well as answers to any maths questions on the resource sheets. We will also indicate misconceptions the children may have.

Focus 23

Def
of 2

Vocabulary

parallel, perpendicular, edge, vertex/vertices, polygon, quadrilateral, parallelogram, rhombus, kite, isosceles, scalene, equilateral, diagonal, right-angled

Resources

RS23 TA for ea
Paper clip for
Ruler for each

ACTIVITY

Pairs

Each child has their own copy of RS23 TA. Children take turns to spin the spinner and read out what it says. (Hold the paper clip to the centre of the spinner with a pencil tip and flick it.)

If the spinner lands on 'perpendicular' (or 'parallel'), they find two dotted lines on the same shape that are perpendicular (or parallel) to each other. They use their ruler and draw over the two dotted lines.

The pair continue until both have drawn over all the lines on their sheet.

Explain why you think those lines are perpendicular to each other.

How could you show those lines are parallel?

Can you draw another line on that shape parallel to one edge?

Other thing

• Children spi
says, then dr

• You can mak
the whole g
and has cou
of the spinne
children who
lines that fit
counters the

ABOUT THE MATHS

Parallel lines go in the same direction and can never meet (even if they
different lengths, start and end at different places and need not be ho

Perpendicular lines are at right angles to each other. They can be of dif
cross and might, or might not, touch.

Focus and learning objective
The Focus number tells you which session in the Teacher's book this activity follows up. The learning objective indicates what children will learn from the activity.

Resources
This is a list of the materials to prepare before your session with the children. There is always a photocopiable resource sheet in the list, which is provided on the opposite page.

We presume you always have pencils and paper to hand, so they are not listed here.

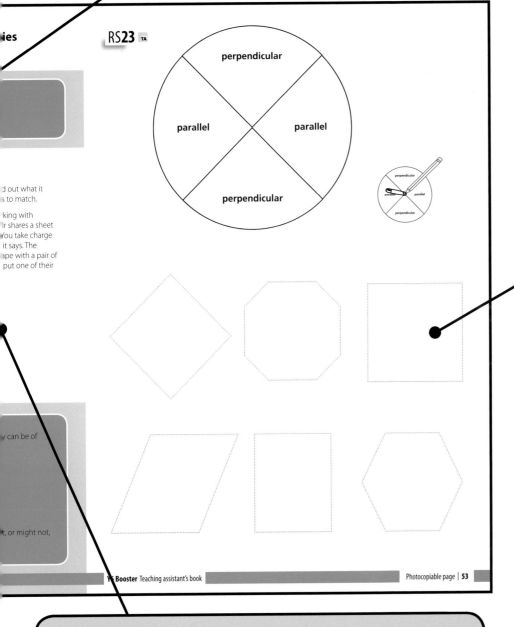

RS**23** ᴛᴀ

Resource sheet
Each session comes with a resource sheet. You give this sheet to the children as it is, cut it up into cards or slips of paper or use the sheet to play a game with. The instructions are always in the 'Activity' section on the opposite page.

There are 10 further resource sheets at the back of the book. These contain materials you might like to use as an additional help: multiplication grids, a place value chart, halving and doubling charts, and so on.

Other things to do
Sometimes there are simpler or different versions of the main activity. Sometimes there are ideas for children who have finished an activity.

Focus 1

Solving problems involving more than one step

Vocabulary

problem, solution, calculate, calculation, operation, answer, method, strategy, explain, reason, predict, estimate, approximate

Resources

RS1 TA for each child (the same resource sheet as RS1 in the Teacher's lesson)
Paper
Calculator for each child

ACTIVITY

Pairs

Pairs share RS1 TA. They choose a different problem than in the teacher's lesson and show as many different ways as they can of solving it. They show all their jottings, including explanations for each step.

A

Each pack holds 24 glue sticks.

There are three packs in a box.

Number of glue sticks in a box: $24 \times 3 = 72$

In six boxes: $72 \times 6 = 432$

Shared between eight classes: $432 \div 8 = 54$

B

There are eight classes.

One pack of 24 glue sticks shared out: $24 \div 8 = 3$

Multiply by 3 (three packs in a box): $3 \times 3 = 9$

Multiply by 6 (six boxes): $9 \times 6 = 54$

C

There are three packs in a box and six boxes.

Total number of packs: $3 \times 6 = 18$

Multiply by 24 (number of glue sticks): $18 \times 24 = 432$

Divide by 8 (number of classes): $432 \div 8 = 54$

Check: $8 \times 54 = 432$

Other things to do

- A more confident child can work with a less confident child. Give the less confident child the worksheet and a pencil; the other child explains to them what to write or draw, but is not allowed to use the pencil themselves.

- Start to solve one of the problems yourself, but stop halfway. As you do this, talk about what you are doing and make sure the child or children are listening. Invite a child to complete the problem you started.

ABOUT THE MATHS

It can help children to draw a diagram to help them solve a problem. For example, to picture the glue stick problem, they might draw the box with three packs of sticks in it. They won't need to draw each stick – writing the number 24 in each pack is enough.

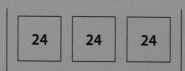

You can offer children calculators to use in solving these problems, but talk with them about what keys they are pressing, and why, and help them understand how to interpret the answer in the display. After all, the calculator doesn't show pound signs or metres, just numbers.

1. A school's order of glue sticks has arrived. There are 24 glue sticks in a pack and three packs in a box. Six boxes were delivered. There are eight classes in the school. How many glue sticks will each class get?

2. A theatre outing costs £20.95 for each adult and £9.45 for each child. How much change does a family of one adult and two children get from a £50 note?

3. At the airport shop, a pair of trainers costs £39.40 without VAT (17.5%). What would they be sold for in the high street with VAT?

4. A class has the computer in their classroom from 1:45 pm until 3:15 pm, four days a week. There are 24 children in the class. How much time can each of them spend on the computer?

5. At playtimes, a school sells apples to the children. Each apple costs 15p. How many apples did they sell this week?

Takings				
Monday	Tuesday	Wednesday	Thursday	Friday
£1.80	£3.45	£5.25	£1.20	£1.65

6. Taslima was having a party. Her mum bought seven 1.5-litre bottles of lemonade. Taslima thought that her fifteen friends would drink about 800 ml each of lemonade. Had Taslima's mum bought enough lemonade?

7. On a school trip there must be one adult to supervise every group of 15 children. All the 218 children in the school went to the seaside. How many adults are needed to supervise?

8. Jenny's bedroom is 3.5 metres long and 3 metres wide. She wants a new carpet, which costs £18.95 a square metre. How much will she have to pay for a new carpet?

9. Seven people share the lottery jackpot of £4 202 576. How much do they each get?

Focus 2

**Solving problems
using a calculator**

Vocabulary

answer, estimate, reasonable, check, calculation,
display

Resources

RS2 TA for each pair
Calculator for each child

ACTIVITY

Pairs

Children work in pairs to find different ways
to get the answer 500 in the calculator display.
They must use at least two different operations
in their calculation when doing this.

For example:

This is not allowed: 250 + 250 = 500

This is allowed: 230 + (9 × 30) = 230 + 270 = 500

Children record the calculations on RS2 TA. When
they have completed 15, they swap with another
pair who will check the calculations with their
own calculator.

Can you use a different operation this time?

Can you invent a three-step calculation?

Tell me a story that goes with this calculation.

Other things to do

• Children invent a real-life story or context for
  one of their calculations. For example: "I had £230,
  then I earned nine pounds a week for 30 weeks.
  £230 + (£9 × 30) = £500." They swap problems
  with another pair to make sure their calculation
  fits the problem.

• Children find as many ways as they can to get
  the answer 5000 in the calculator display.

ABOUT THE MATHS

There are unlimited calculations with 500 as the answer, so finding 15 of them should not be too hard.
Ask children to use different operations: multiplication and division as well as addition and subtraction.

When calculations include brackets, this means 'do the bit inside the bracket first':

$100 × (30 − 25) = 100 × 5 = 500$

If children write down the calculation without brackets

$100 × 30 − 25 =$

the pair checking may do the operations in order and not get the answer intended:

$100 × 30 − 25 = 3000 − 25 = 2975$

10 | **Mathematics Accomplished: Y6 Booster** Teaching assistant's book

500

Find 15 different ways to make 500.

You must use at least two different operations in each calculation:

This is not OK:
250 + 250

This is OK:
230 + (9 × 30)

_____   _____   _____

_____   _____   _____

_____   _____   _____

_____   _____   _____

_____   _____   _____

Calculations

Checked by

Focus 3

Using positive and negative numbers in context

Vocabulary

positive, negative, compare, order, ascending, descending, subtract, difference, minus

Resources

RS3 TA for each child
Calculator for each child

ACTIVITY

Individuals and pairs

Each child has a copy of RS3 TA. They find out how many jumps there are between each pair of numbers.

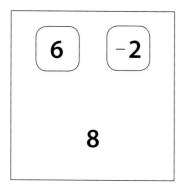

They then invent their own problems for their partner to tackle.

Can you show us on the number line how you worked that out?

What is a quick way of counting the first set of jumps? How does zero help?

Do you think a jump from -7 to -1 will be more or less than 7 jumps? Why do you think that?

Other things to do

- Children invent their own problems using numbers in a wider range: for example, −15 to 15. They can tackle these together with their partner.

- Challenge children to do some of the calculations without looking at the number line.

ABOUT THE MATHS

The number line at the bottom of the worksheet is an important tool for children in working out these problems. For example, to find the difference (the number of steps) between 6 and −2, children can simply count the steps on the line.

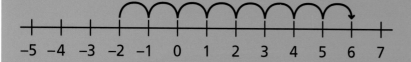

The difference between 6 and −2 is 8.

A quicker way is to make two jumps and add the number of steps in each.

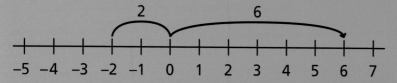

Find out how many jumps there are between each pair of numbers.
Use the number line below to help.

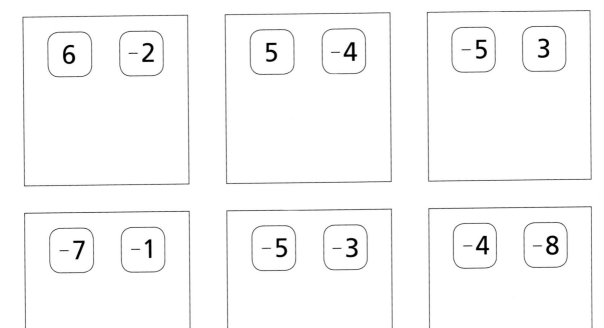

6	−2
5	−4
−5	3
−7	−1
−5	−3
−4	−8

Now make up some for your partner to do.

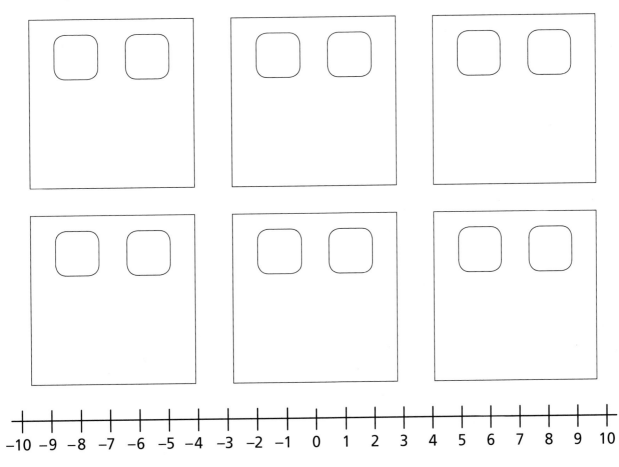

Focus 4

Using decimal notation for tenths and hundredths

Vocabulary

place value, digit, numeral, partition, decimal point, decimal place, tens, units, tenths, hundredths

Resources

RS4 TA for each pair

0–9 dice for each group

50 counters in two different colours for each group

ACTIVITY

Groups of four

Children play as two pairs. The aim of this game is to make lines of four counters in your own team's colour, sideways, up and down or diagonally.

Groups share RS4 TA. The teams take turns to roll the dice twice. If they roll a double, they must roll the dice again.

Team A: One child decides where to place a counter (on any number that is between the two dice numbers) and instructs their partner where to place the counter by reading out the number. Help children use the number line on the board when thinking about what numbers are allowed.

If a team completes a line of four counters in their own colour, they take them off the board and score a point. Keep playing until one team has scored 5 points.

Roughly where on this line does 7.18 belong? So is it between 3 and 8?

What two dice numbers could 7.098 be between?

What pair of numbers do you hope to get next?

Other things to do

- Look with children at any row or column of 7 numbers on the board and ask:

  Which is the largest/smallest number in this row/column?

- Choose any two whole consecutive numbers from the number line, such as 4 and 5. Children show you any number on the board which lies between those two numbers: for example, 4.43.

ABOUT THE MATHS

When working in pairs, ensure that one child reads out the number they want so their partner can place a counter on this number on the board. This helps them practise reading decimal numbers. They should read out the numbers correctly: for example, 2.84, not 'two point eighty-four'.

In this game, the actual value of the decimal part of the number doesn't matter. If a player rolls a 4 and a 5, they can put a counter on 4.41 or 4.46. Both numbers are equally valid. Children need to focus on the units number to the left of the decimal point. They can look at the 4 units and know that as the number is a decimal – that is, slightly more than 4, but not 5.

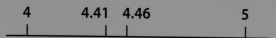

4.41 and 4.46 are both between 4 and 5.

1.12	1.3	1.06	1.003	1.9	2.85	2.03
2.72	2.05	2.09	3.9	3.22	3.54	3.24
3.32	4.41	4.46	4.43	4.01	4.4	5.02
5.91	5.26	5.55	5.74	5.61	6.91	6.6
6.05	6.84	6.71	7.02	7.6	7.18	7.14
7.9	7.09	8.7	8.01	8.11	8.19	8.5
8.12	9.24	9.99	9.02	9.1	9.25	9.01

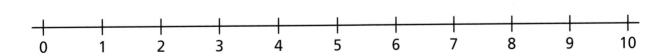

Focus **5**

Comparing decimals

Vocabulary

place value, digit, numeral, partition, decimal point, decimal place, units, tenths, hundredths, thousandths, compare, order, greater than, less than

Resources

RS5 TA for each child

Coin

0–9 dice for each pair

ACTIVITY

Pairs

Children play 'Decimal roll' in pairs. First, they toss a coin to decide: heads for larger, tails for smaller. Each child then rolls the 0–9 dice three times and uses their results as a tenths, hundredths and thousandths digit. They each arrange their digits with zeros in the units place. The child who makes the larger (or smaller) number wins a point.

Coin: heads (larger)

Player 1

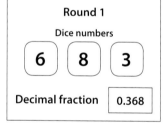

Player 2

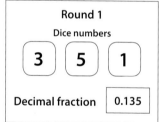

The player with more points after 10 rounds is the winner.

What is a decimal fraction?

Can you explain why 0.368 is larger than 0.359?

Where will you put that zero in your decimal fraction?

Other things to do

- Children roll the dice to make decimal numbers such as 3.72 (U.th).

- Children roll the dice four times and choose three of them to make their number. For example, a child gets 2 6 1 9 and knows they are aiming for 'smaller', so they ignore the 9 and make the number 0.126.

ABOUT THE MATHS

Help children compare numbers by looking at the digits which are worth the most: 0.911 is larger than 0.899, because nine tenths (the 9 in 0.911) is worth more than eight tenths (the 8 is 0.899). The hundredths and thousandths digits can be ignored.

But with two numbers such as 0.409 and 0.411, you need to look at the hundredths digits, because the tenths are the same. Here, 0.411 is larger than 0.409, because 1 is larger than 0. A decimal fraction is a number less than 1, shown using a 0 and a decimal point.

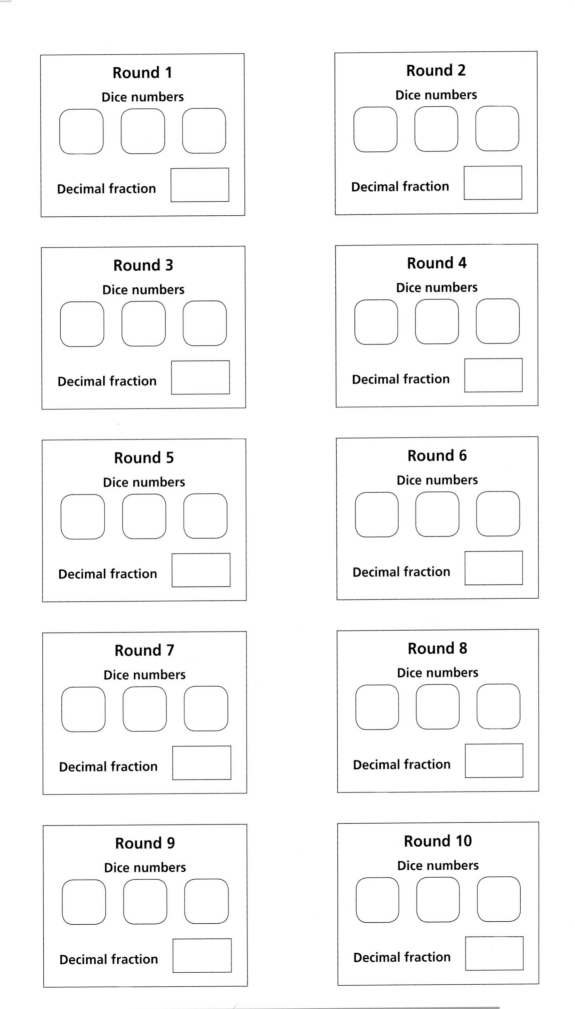

Round 1

Dice numbers

Decimal fraction

Round 2

Dice numbers

Decimal fraction

Round 3

Dice numbers

Decimal fraction

Round 4

Dice numbers

Decimal fraction

Round 5

Dice numbers

Decimal fraction

Round 6

Dice numbers

Decimal fraction

Round 7

Dice numbers

Decimal fraction

Round 8

Dice numbers

Decimal fraction

Round 9

Dice numbers

Decimal fraction

Round 10

Dice numbers

Decimal fraction

Focus 6

Vocabulary

digit, tens, units, tenths, hundredths, thousandths, compare, greater than, less than, order, round, estimate, approximate, approximately

Resources

RS6 TA for each child
0–9 dice for each child

ACTIVITY

Pairs or groups of four

Each child has their own copy of RS6 TA. Children take turns to roll the dice three times and use the digits to make a decimal number with two decimal places. (They can use the digits in any order.)

8.13 or 3.18 or 8.31 or 1.38 or …

They round their decimal number to the nearest whole number and write the decimal on the balloon showing that whole number.

8.13 rounds to 8.

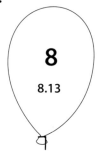

8

8.13

The first child to write a number on each of their balloons is the winner.

Can someone explain how to round a number where the last digit is 5?

What numbers do you need to roll to make a decimal number for your last balloon?

Which numbers do you take notice of when you want to round to the nearest whole number?

Other things to do

- Help children locate their numbers on the number line on RS6 TA. For example, 8.13 comes between 8 and 9, but closer to the 8.

- Children roll the dice twice and make a number with just one decimal place. They then round this to the nearest whole number.

6.3 rounds to 6.

ABOUT THE MATHS

The rules for rounding a number with one or two decimal places to the nearest whole number are the same:

- If the tenths digit is 0, 1, 2, 3 or 4, round down.

- If the tenths digit is 5, 6, 7, 8 or 9, round up.

- You can ignore the hundredths digit.

Units	tenths	hundredths	
8 .	3		rounds down to 8
2 .	9	1	rounds up to 3

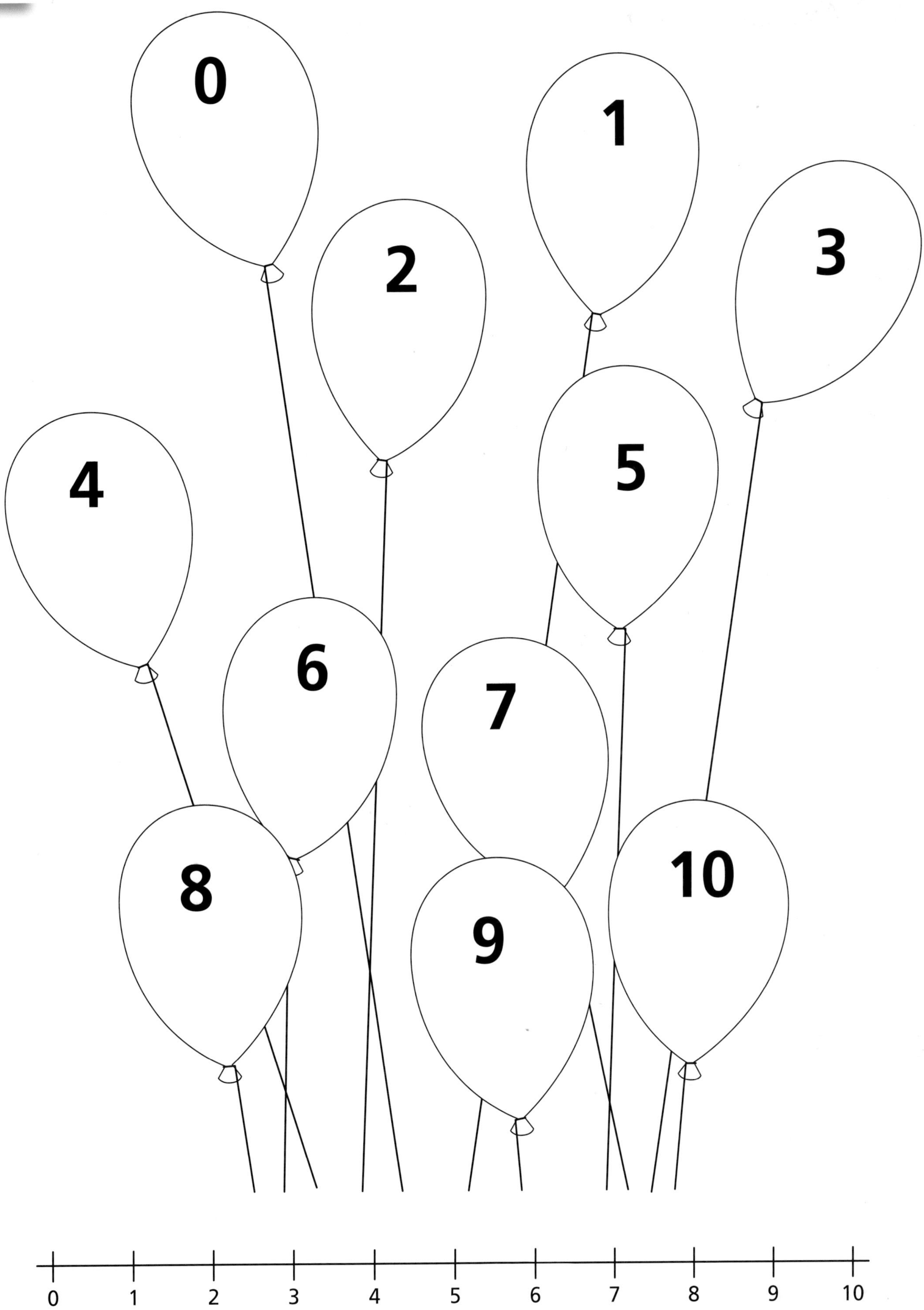
0
1
2
3
4
5
6
7
8
9
10
0 1 2 3 4 5 6 7 8 9 10

Focus 7

Ordering fractions

Vocabulary

whole, part, equal parts, numerator, denominator, order, equivalent, improper fractions, mixed numbers, simplest form

Resources

RS7 TA for each pair or group

A3 sheet of paper for each pair

ACTIVITY

Pairs

Children shuffle the fraction cards cut from RS7 TA and place them in a pile, face down. They draw a long line on a sheet of A3 paper turned sideways, marking one end 0 and the other end 1.

They take turns to pick a fraction card and place it at approximately the correct position on the line.

Each fraction card is equivalent to three others. A player who completes a column of four equivalent fractions wins a point. When all the fraction cards are placed on the line, the child with the most points wins the game.

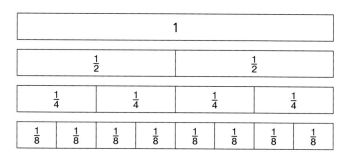

How do you know that $\frac{9}{12}$ is the same as $\frac{3}{4}$?

These four cards are all the same value. Which of the fractions cannot be simplified further?

Other things to do

- Children shuffle the fraction cards and place them in a pile, face down. They both pick a card and whoever has the card of higher value wins both cards.

ABOUT THE MATHS

You may need to help children check whether the fractions are the same by simplifying them. Divide both the top part of the fraction (the numerator) and the bottom part (the denominator) by the same number.

For example:

To simplify $\frac{6}{10}$, divide the 6 and the 10 by 2. $\frac{6}{10} \rightarrow \frac{3}{5}$

To simplify $\frac{4}{12}$, divide the 4 and the 12 by 4. $\frac{4}{12} \rightarrow \frac{1}{3}$

A number line with each fraction in the correct position will look like this:

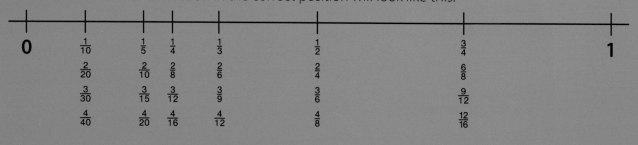

$\dfrac{1}{2}$	$\dfrac{2}{4}$	$\dfrac{3}{6}$	$\dfrac{4}{8}$	$\dfrac{1}{3}$
$\dfrac{2}{6}$	$\dfrac{3}{9}$	$\dfrac{4}{12}$	$\dfrac{1}{10}$	$\dfrac{2}{20}$
$\dfrac{3}{30}$	$\dfrac{4}{40}$	$\dfrac{1}{5}$	$\dfrac{2}{10}$	$\dfrac{3}{15}$
$\dfrac{4}{20}$	$\dfrac{1}{4}$	$\dfrac{2}{8}$	$\dfrac{3}{12}$	$\dfrac{4}{16}$
$\dfrac{3}{4}$	$\dfrac{6}{8}$	$\dfrac{9}{12}$	$\dfrac{12}{16}$	

Focus 8

Recognising equivalence between fractions, decimals and percentages

Vocabulary

percentage, decimal, fraction, convert, equivalent, numerator, denominator

Resources

Sets of fraction cards cut from top half of RS8 TA

Percentage and decimal cards cut from bottom half of RS8 TA

Counters

ACTIVITY

Group

Children choose any nine fraction cards and form an array, as on a Bingo board. Shuffle the percentage and decimals cards and put them in a pile on the table, face down. Take one of these percentage cards and read it out.

The children then look at their cards: if they have an equivalent fraction, they say what it is and cover it with a counter. Some of your cards will have more than one equivalent (for example, 50% is the same as $\frac{1}{2}$, $\frac{50}{100}$ and also 0.5). Children may only cover one of these equivalents, even if they spot more than one. The first player to cover all their numbers says 'Bingo' and wins the game. Continue until everybody has covered their board.

Can you explain why a half and fifty hundredths are both 50%?

Can you tell us how to turn 0.75 into a fraction?

Can you give another fraction that is also the same as 20%?

Other things to do

- **Tough Bingo**
  The first child to say that they have the fraction to match your card wins a counter, but any other players with that fraction don't. Replace cards you have used in the pack so that children have a second chance of getting a counter for that number.

- Help children sort out the fraction, decimal and percentage cards so that ones of the same value are together:

$\frac{1}{2}$	$\frac{50}{100}$	50%	0.5
$\frac{1}{4}$	$\frac{25}{100}$	0.25	
$\frac{1}{5}$	$\frac{20}{100}$	20%	0.2
$\frac{1}{10}$	10%	0.1	
$\frac{3}{4}$	$\frac{75}{100}$	75%	0.75
$\frac{1}{100}$	1%		
$\frac{2}{100}$	2%		
$\frac{1}{20}$	5%		

ABOUT THE MATHS

To change a fraction to a decimal or a percentage, you need to find the equivalent fraction which is a number of hundredths.

$\frac{3}{4}$ (multiply top and bottom by 25) $\rightarrow \frac{75}{100}$ $\frac{75}{100}$ **is the same as 0.75 or 75%.**

To change a percentage into a decimal, write the percentage number as hundredths.

20% $\rightarrow \frac{20}{100} \rightarrow$ 0.2

To change a percentage into a fraction, write the percentage number as hundredths, then simplify the fraction if you can.

20% $\rightarrow \frac{20}{100}$ (divide top and bottom by 10) $\rightarrow \frac{2}{10}$ (divide top and bottom by 2) $\rightarrow \frac{1}{5}$

$\frac{1}{2}$	$\frac{3}{4}$	$\frac{1}{5}$	$\frac{1}{100}$
$\frac{1}{10}$	$\frac{25}{100}$	$\frac{50}{100}$	$\frac{20}{100}$
$\frac{1}{4}$	$\frac{75}{100}$	$\frac{2}{100}$	$\frac{1}{20}$

0.1	0.75	0.5	0.2
0.25	20%	10%	50%
5%	75%	1%	2%

Focus 9

Solving simple scaling problems

Vocabulary

half, quarter, amount, quantity, table, scale up/down

Resources

RS9 TA for each child

ACTIVITY

Pairs

Each child works with their own copy of RS9 TA, but they work with a partner so they have someone to talk with.

When you increase the amount of sugar, how can you make sure you increase the amount of fairy tears in the same proportion?

If you know the recipe for one dose, how do you work out the ingredients for five doses?

If you know how many worms you need for two doses, how would you work out the number of worms you need for six doses?

Other things to do

- Children could invent their own ingredients entirely. The list on RS7 TA is only there to provide ideas.

- Confident children can put more than three ingredients in their recipes.

ABOUT THE MATHS

1. To make 5 doses of the wart removal recipe involves multiplying each amount by 5.

1 dose	5 doses
4 toad eggs	20 toad eggs
2 worms	10 worms
6 powdered vampire teeth	30 powdered vampire teeth

2. 24 vampire teeth ÷ 6 (the number you need for each dose) = 4 doses

    Check by multiplying: $6 \times 4 = 24$

    Multiply each ingredient by 4:

    4 toad eggs $\times$ 4 = 16 toad eggs

    2 worms $\times$ 4 = 8 worms

    6 vampire teeth $\times$ 4 = 24 vampire teeth

    Those ingredients mashed up will be enough to make 4 doses.

3. With the sweet dreams recipe, children need to divide each ingredient by 10 to find the amount for 1 dose. They then multiply these amounts by 6 to get the amounts for 6 doses.

10 doses	1 dose	6 doses
1000 fairy tears	100 fairy tears	600 fairy tears
40 violet petals	4 violet petals	24 violet petals
5 teaspoons sugar	$\frac{1}{2}$ teaspoon sugar	3 teaspoons sugar

To remove a wart

1 dose

4 toad eggs
2 worms
6 powdered vampire
   teeth

Mash up the ingredients and take in water morning and evening.

To have sweet dreams

enough for 10 doses

1000 fairy tears
40 violet petals
5 teaspoons sugar

Gently heat ingredients until sugar is melted. Leave to cool and bottle. Keep in fridge.

1. Work out how much of each ingredient you need to make your recipe for 5 doses of wart remover.

2. Work out how much of each ingredient you need to remove warts if you have 24 powdered vampire teeth in your mixture.

   How many warts can you remove?

3. Look at the sweet dreams recipe. How much of each ingredient do you need for 1 dose?

   And for 6 doses?

Focus 10

Using known facts to multiply and divide decimals

Vocabulary

method, strategy, explain, predict, digit, decimal point, decimal place, multiply, divide

Resources

RS10 TA for each child

Two 0–9 dice for each group

Calculator for each child (optional)

ACTIVITY

Groups of 3 or 4

Each child has their own copy of RS10 TA. Playing in small groups, children take turns to roll the dice and write the two dice numbers in the diamonds on their sheet, in whichever order they prefer.

Round 1

$0 . \langle 4 \rangle \times \langle 9 \rangle = \boxed{}$

or

Round 1

$0 . \langle 9 \rangle \times \langle 4 \rangle = \boxed{}$

They work out the answer and write it in the box.

The child with the largest product is the winner of that round.

Children play 10 rounds.

Can you explain to your partner how you know where to put the decimal point?

Show someone else how you did that calculation.

If three sixes are eighteen, what do you think three nought point sixes are?

Other things to do

- Children check their answers with a calculator.

- Children draw up and complete tables like this to help them work out multiplication facts involving decimals.

×	1	2	3	4	5	6	7	8	9	10
5	5	10	15							
0.5	0.5	1.0	1.5							

ABOUT THE MATHS

In the teacher's session, children used their knowledge of the multiplication tables to work out multiplication facts involving decimals.

For example, if 4×7 is 28, then 4×0.7 is 2.8, and 0.4×7 is also 2.8.

In both 4×0.7 or 0.4×7, one digit has moved to the right one place compared with 4×7. So the answers are the same as for 4×7, with each digit also moved one place to the right.

	Tens	Units	tenths
4×7	2	8	
4×0.7		2	8
0.4×7		2	8

26 | **Mathematics Accomplished: Y6 Booster** Teaching assistant's book

Round 1 0 . ◇ × ◇ = ▭

Round 2 0 . ◇ × ◇ = ▭

Round 3 0 . ◇ × ◇ = ▭

Round 4 0 . ◇ × ◇ = ▭

Round 5 0 . ◇ × ◇ = ▭

Round 6 0 . ◇ × ◇ = ▭

Round 7 0 . ◇ × ◇ = ▭

Round 8 0 . ◇ × ◇ = ▭

Round 9 0 . ◇ × ◇ = ▭

Round 10 0 . ◇ × ◇ = ▭

Focus 11

Vocabulary

square, square, number, pattern, array, predict

Resources

Cards cut from RS11 TA for each pair

ACTIVITY

Pairs

Pairs of children spread out the cards from RS11 TA on the table, face down.

They take turns to turn over one rectangular card and one square card. If the cards match, the child keeps both cards. If the cards do not match, the child puts them back where they were, face down.

The game continues until all the equivalent pairs have been collected. The child with the most cards is the winner.

This answer card says 100. What calculation do we need to find to go with it?

We're looking for six sixes. Do you think the answer will be even or odd?

What's a quick way of multiplying by 11 by 11?

Other things to do

- Spread out the rectangular cards on the table, face down. Place the set of square cards face down in a pile. Turn over one of the square cards face up at a time and ask the children quickly to find the card that goes with it.

| $4 \times 4 =$ | | 16 |

- Children arrange nine of the rectangular cards as a 3 × 3 Bingo board. Call out answers from the square cards and children cover the related multiplication with a counter. The first child to cover all of their cards with a counter wins the game.

ABOUT THE MATHS

The square 'answer' cards show the square numbers to 12 × 12: 1, 4, 9, 16, 25, 36, 49, 64, 81, 100, 121, 144.

When all the rectangular cards are matched up with the answer cards, you will get the following:

$1 \times 1 = 1$	$7 \times 7 = 49$
$2 \times 2 = 4$	$8 \times 8 = 64$
$3 \times 3 = 9$	$9 \times 9 = 81$
$4 \times 4 = 16$	$10 \times 10 = 100$
$5 \times 5 = 25$	$11 \times 11 = 121$
$6 \times 6 = 36$	$12 \times 12 = 144$

$1 \times 1 =$

$2 \times 2 =$

$3 \times 3 =$

$4 \times 4 =$

$5 \times 5 =$

$6 \times 6 =$

$7 \times 7 =$

$8 \times 8 =$

$9 \times 9 =$

$10 \times 10 =$

$11 \times 11 =$

$12 \times 12 =$

1	4	9	16
25	36	49	64
81	100	121	144

Focus 12

Vocabulary

multiple, factor, divisor, divisible, divisibility, prime, prime factor

Resources

RS12 TA for each child

Coloured pencils

0–9 digit cards

ACTIVITY

Individuals or pairs

Children complete RS12 TA and then play the game, taking turns to turn over two digit cards and using these to make a two-digit number. If the number is not already crossed out on the grid, they write it in one of the empty squares below the grid and circle it on the grid. The winner is the first to fill the 10 empty boxes.

What caused you to cross out 8? and 15?

Can you explain why you haven't crossed out 17 yet? Do you think you will?

What do you notice about the numbers you have circled?

Other things to do

• Give children a couple of minutes to look carefully at their completed worksheets, then remove them. Ask questions about whether numbers are prime or not: "Is 31 a prime number? Is 32?" Children win a counter for each correct answer.

• Help children check that prime numbers really are prime by trying to divide them, using a calculator.

If a larger number is divisible by another number, the calculator will show a whole number answer.

For example: $8 \div 2$ gives 4, which means 8 is divisible by 4.

But if it isn't, there will be a decimal answer, meaning there is a remainder.

For example: $13 \div 2$ gives 6.5, which means 13 is not divisible by 2.

Children will find that whatever they divide a prime number by, they will get a decimal answer – unless, of course, they divide the number by itself or 1.

For example: $11 \div 1$ gives 11, and $11 \div 11$ gives 1.

ABOUT THE MATHS

Once children have crossed out the numbers as instructed, they will be left with all the prime numbers under 100. Prime numbers such as 2, 11 and 19 are numbers that cannot be divided up into smaller whole numbers. The only factors of a prime number are the number itself, and 1.

The factors of a number are those numbers which divide exactly into the number. For example: 24 has these factors: 1, 2, 3, 4, 6, 8, 12, 24. But the factors of 11, which is prime, are just 1 and 11.

RS12 TA

Look at the 1–100 number square below.

- Cross out 1.
- Cross out all the multiples of 2, but not 2.
- Cross out all the multiples of 3, but not 3.
- Cross out all the multiples of 5, but not 5.
- Cross out all the multiples of 7, but not 7.

1	2	3	4	5	6	7	8	9	10
11	12	13	14	15	16	17	18	19	20
21	22	23	24	25	26	27	28	29	30
31	32	33	34	35	36	37	38	39	40
41	42	43	44	45	46	47	48	49	50
51	52	53	54	55	56	57	58	59	60
61	62	63	64	65	66	67	68	69	70
71	72	73	74	75	76	77	78	79	80
81	82	83	84	85	86	87	88	89	90
91	92	93	94	95	96	97	98	99	100

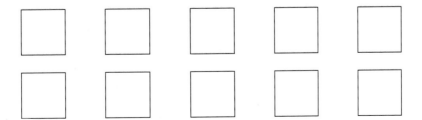

Focus 13

Vocabulary

digit, define/definition, multiple, multiply, factor, divisible, multiplication, product, table, record

Resources

RS13 TA for each pair

ACTIVITY

Pairs

Children choose a factor from those listed and complete the sentence 'These numbers are divisible by …'. For example:

These numbers are divisible by 3

They then think up nine numbers to write in the box that fit the description, but they add *one* number which does not.

For example, here, 29 does not belong as it is not divisible by 3.

These numbers are divisible by 3	12 21 30 99 9372 120 351 29 540 60

They swap sheets with another pair who spot the deliberate mistake in each set of numbers.

Can you explain why you chose 484 as a number divisible by 4?

What's a quick way of inventing a 3-digit number divisible by 3?

540 is divisible by 10. What else is it divisible by?

Other things to do

- Help children check whether a larger number is divisible by a smaller number, using a calculator. If the larger number is divisible by the other number, the calculator will show a whole number answer.

  For example: $64 \div 4$ is 16, so 64 is divisible by 4.

  If it isn't, there will be a decimal answer, meaning there is a remainder.

  For example: $69 \div 8$ gives 8.625, so 69 is not divisible by 8.

- Say some numbers and ask children to state which, if any, numbers they are certain will divide exactly into your number.

ABOUT THE MATHS

A number is divisible by another number if it divides by that number exactly, with no remainder. For example: 42 is divisible by 3, because $42 \div 3 = 14$.

You can check which small numbers a larger number is divisible by, using the tests below.

2	The last digit is 0, 2, 4, 6 or 8.		8	The last three digits are divisible by 8.
3	The sum of the digits is divisible by 3.		9	The sum of the digits is divisible by 9.
4	The last two digits are divisible by 4.		10	The last digit is 0.
5	The last digit is 0 or 5.		25	The last two digits are 00, 25, 50 or 75.
6	The number is even and divisible by 3.		100	The last two digits are 00.

2	The last digit is 0, 2, 4, 6 or 8.
3	The sum of the digits is divisible by 3.
4	The last two digits are divisible by 4.
5	The last digit is 0 or 5.
6	The number is even and divisible by 3.
8	The last three digits are divisible by 8.
9	The sum of the digits is divisible by 9.
10	The last digit is 0.
25	The last two digits are 00, 25, 50 or 75.
100	The last two digits are 00.

These numbers are divisible by	
These numbers are divisible by	
These numbers are divisible by	
These numbers are divisible by	

Focus 14

Learning how to check

Vocabulary

inverse, operation, reverse, order, sum, product, strategy, divisible by, multiple of

Resources

RS14 TA for each child

Two sets of 0–9 digit cards for each pair

Calculator for each child

ACTIVITY

Individuals and pairs

Each child has their own copy of RS14 TA. They shuffle two sets of 0–9 digit cards and place them in a pile on the table, face down. They each take four cards and use the digits to make two two-digit numbers, writing these down as a multiplication calculation on RS14 TA. They do this five times each, creating calculations for their partner to complete.

The pair swaps sheets and estimates the answer to each calculation they have been given. They then work out the answers accurately, using a calculator. Encourage children to congratulate themselves if their answers are within 300 of their estimate.

Calculation	Estimate	Calculator answer
19×34	$20 \times 30 = 600$	646
26×73	$20 \times 70 = 1400$	1898
38×56	$40 \times 60 = 2400$	2128

How can you round those numbers to make a quick estimate?

Does that estimate seem about the right size? Why do you think it is?

Other things to do

- Children take four cards and use the digits to make a three-digit number and a single digit to multiply: for example, 152×7.

- Help children use what they know of their multiplication tables to work out the final digit as part of their estimation.

Calculation	Estimate	Calculator answer
29×32	900; 9 × 2 is 18, so the number will end in 8.	928
51×99	5000; 1 × 9 is 9, so the number will end in 9.	5049

ABOUT THE MATHS

In the teacher's session, children used different ways of checking calculations, such as:

- using the inverse operation (to check $35 \times 20 = 700$, do $700 \div 20$ and expect the answer 35)

- doing an equivalent calculation (to check $35 \times 20 = 700$, do 350×2)

- looking for patterns of numbers (to check $35 \times 20 = 700$, check that the answer ends in zero)

- estimating whether the answer is about the right size (to check $35 \times 20 = 700$, expect an answer less than 800, but more than 600)

It is the last of these methods that children are practising in this activity: in this case, they estimate the size of the answer to expect before working out the actual answer with a calculator.

Calculation	Estimate	Calculator answer
☐☐ × ☐☐		
☐☐ × ☐☐		
☐☐ × ☐☐		
☐☐ × ☐☐		
☐☐ × ☐☐		

Focus 15

Finding fractions and percentages of whole-number quantities

Vocabulary

percentage, per cent (%), fraction, numerator, denominator, equivalent, simplify, cancel, divide

Resources

RS15 TA for each child

ACTIVITY

Individuals

Children write in under the lines as many different percentages as they can.

On the last line, children choose an amount for themselves to represent 100%.

Show me a way to find a percentage of £200.

Would you rather have 30% of £200 or 50% of £100?

Write a question for everyone to solve about your last line of percentages.

Other things to do

- Children use a calculator to check their working out for 50% (divide by 2) and 10% (divide by 10) of the amounts.

- Ask children if it would help if they turned the £4 (line 2) into 400p. If not, make sure they know that 10% of £4 is 40p.

ABOUT THE MATHS

Children should begin by finding simple percentages such as 50% (half) and 10% (one tenth). From here, they can work out 20% (by doubling 10%) and 5% (by halving 10%).

10% of £200 is £20.   **So 20% of £200 is £40.**

10% of £200 is £20.   **So 5% of £200 is £10.**

They can also work out all the other multiples of 10%.

10% of £200 is £20.
So 30% of £200 is £20 × 3 or £60.

Other possibilities: Find 1% by dividing 10% by 10; find 15% by adding 5% and 10%.

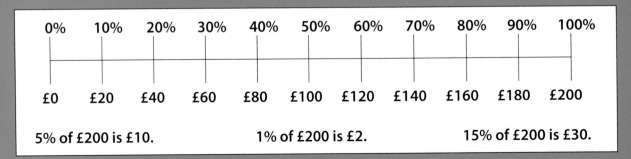

5% of £200 is £10.   1% of £200 is £2.   15% of £200 is £30.

Use the 0% to 100% number lines to help you find as many different percentages as you can.

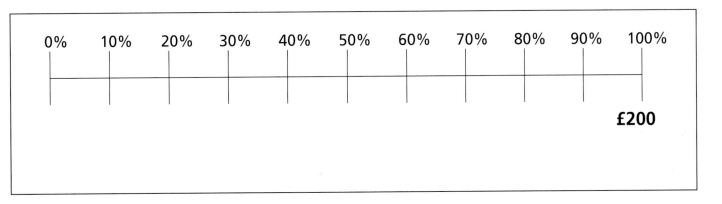

£200

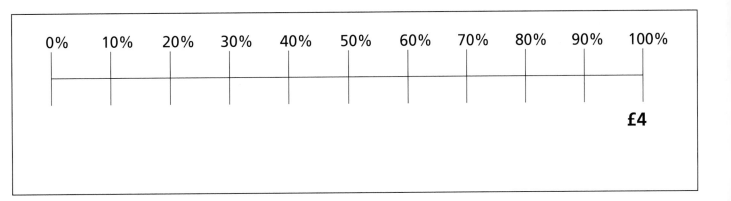

£4

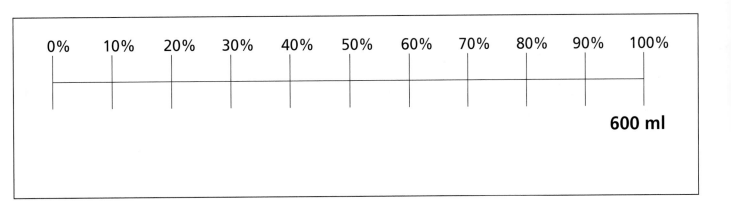

600 ml

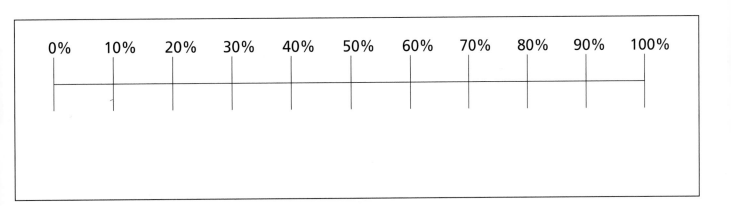

Focus 16

Adding and subtracting decimal numbers mentally

Vocabulary

decimal, decimal point, decimal fraction, sum, difference, answer, solution, method

Resources

RS16 TA for each child

0–9 dice for each child

Calculator for each child (optional)

ACTIVITY

Individuals, then pairs

Each child chooses any number under 10 with one decimal place and writes it down on their copy of RS16 TA. They roll a 0–9 dice and work out what to add to, or subtract from, their decimal number to make the dice number.

Decimal number:	4.5
Dice number:	2
Calculation:	4.5 − 2.5 = 2

Children do this 10 times. They then swap work with another child and check each other's work.

Supposing you start with 4.5 and your dice number is 2. How do you work it out? How does that help you?

Tell your partner how you worked that out.

You don't think that is correct? Then explain to your partner how you would do it.

Other things to do

- Choose a number under 10 with one decimal place, then work out what to add to that number to make 10. For example, $3.5 + ? = 10$

- Children work in pairs. Each child chooses any number under 10 with one decimal place and writes it down. Together they work out what to add to, or subtract from, one of their numbers to reach the other number.

ABOUT THE MATHS

Help children draw empty number lines (as in the main activity with the teacher) to work out their calculation in stages.

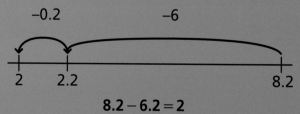

$$8.2 − 6.2 = 2$$

They can also use a calculator to check their calculations. For example, a child thinks that $4.5 − 2.5$ is 2, but is not quite sure. Help them do the calculation on the calculator as a check.

Decimal number
(less than 10):

Dice number:

Calculation:

Decimal number
(less than 10):

Dice number:

Calculation:

Decimal number
(less than 10):

Dice number:

Calculation:

Decimal number
(less than 10):

Dice number:

Calculation:

Decimal number
(less than 10):

Dice number:

Calculation:

Decimal number
(less than 10):

Dice number:

Calculation:

Decimal number
(less than 10):

Dice number:

Calculation:

Decimal number
(less than 10):

Dice number:

Calculation:

Decimal number
(less than 10):

Dice number:

Calculation:

Decimal number
(less than 10):

Dice number:

Calculation:

FOCUS 17

Using strategies for halving whole and decimal numbers

Vocabulary

half/halving, double, strategy, partitioning, digit, decimal point, decimal place, estimate, approximate

Resources

RS17 TA for each child
0–9 dice for each pair

ACTIVITY

Pairs

Each child has a copy of RS17 TA. Children agree a whole-number target under 10, such as 8, and write it on RS17 TA. One child rolls the 0–9 dice twice, and they both record these numbers, too.

Each child in secret decides how to arrange the dice digits to make a number with one decimal place.

You could make 4.6 or 6.4.

Children then decide in secret whether to:

* double their number

* halve it or

* keep it as it is.

They compare their answers, and the child with the answer closer to the target number wins a point. If both are equally close, they each score a point.

Children play 15 rounds. The child who wins more points is the overall winner.

Your target number is small. How are you going to make your answer small?

What decimal number can you make that you can double to get near your target number?

Can you make a bigger number and halve it? Or a smaller number and double it?

Other things to do

* Children use a number line to check which number is closer to the whole-number target: "Is 9.2 or 6.6 closer to 8?"

* Instead of working with decimal numbers, children use the dice numbers to make two-digit numbers such as 45 or 19. They aim to get close to a whole-number target under 100, such as 50.

Whole-number target	Dice numbers	Decimal number	Double, halve or keep it as it is?	Score
8	4 6	4.6	Double 4.6 × 2 = 9.2	1

ABOUT THE MATHS

The main focus of the teacher's session is halving. If you can halve a number such as 24 or 25, you can also halve the decimal numbers which are a tenth of the size.

Half of 24 is 12, so half of 2.4 is 1.2.

Half of 43 is 21.5, so half of 4.3 is 2.15.

When halving decimal numbers, it is helpful to work with whole numbers, then add the decimal point as necessary.

Whole-number target	Dice numbers	Decimal number	Double, halve or keep it as it is?	Score

Focus 18

Using doubling and halving to work out unknown multiplication facts

Vocabulary

number facts, double/doubling, strategy, multiplication

Resources

RS18 TA for each pair

Two different-coloured pencils for each pair

ACTIVITY

Pairs

Children take turns to choose a number from Group A and one from Group B, then multiply them together. Provide paper for them to work out the answer if they need it. They find the answer number on the large grid and draw a cross through that number with their coloured pencil.

If there is already a cross through that number or it does not appear on the grid, they miss that turn.

The first child to have a row, column or diagonal of four numbers in their colour is the winner.

Which number do you want to go for? What could you multiply together to get that number?

What's an easy way of getting a whole number answer?

This number ends in point five. What could you multiply together to get that?

Other things to do

- The child who is waiting for their turn can check their partner's calculation on a calculator. Don't let them reveal the answer until the child working out the multiplication has finished.

- Children talk to their partner about how they might do the multiplication before they actually tackle it.

ABOUT THE MATHS

Children will use a variety of methods for working out the multiplications. They may work out a multiplication as if there were no decimal point involved, then put it back in.

35×3 is 105, so 3.5×3 is 10.5.

They may also use partitioning by breaking the number up into smaller parts and doing a bit at a time.

$2.7 \times 4 = ?$ $2 \times 4 = 8$ and $0.7 \times 4 = 2.8$

So $2.7 \times 4 = 8 + 2.8 = 10.8$.

Or they may use repeated doubling.

$4.6 \times 8 = ?$ $4.6 \times 2 = 9.2$

$4.6 \times 4 = 18.4$

$4.6 \times 8 = 36.8$

Group A	
3.5	4.3
8.2	2.7
3.4	4.6

Group B	
20	8
10	6
3	4

43	12.9	24.6	8.1	68	13.8
14	164	32.8	10.8	13.6	21
18.4	25.8	21.6	16.2	20.4	27.6
28	34.4	65.6	49.2	27.2	36.8
35	10.5	82	27	34	46
70	86	17.2	54	10.2	92

Focus 19

Using a pencil-and-paper method for addition and subtraction

Vocabulary

method, explain, reason, place value, digit, decimal point, decimal place, tens, units, tenths, hundredths, thousandths, estimate, approximate

Resources

RS19 TA for each child
0–9 digit cards for each pair

ACTIVITY

Pairs

Each child has their own copy of RS19 TA. Children shuffle the cards and deal three each. Each child leaves their cards face up. One player decides whether, in this round, they are aiming for the highest or lowest answer. Both players circle the appropriate arrow. They arrange their cards to make a decimal number which they write in the boxes on RS19 TA.

Each child then completes the calculation and writes the answer in the box. Children compare answers. The winner of that round (that is, the player with the highest or lowest answer) draws a circle around the thumbs up; the loser draws a circle around the thumbs down. Children shuffle the cards again and play another round. They continue for eight rounds. The child with more thumbs up is the overall winner.

What number will you make to give you the largest total?

What number will you make so that you can subtract the smallest amount?

Can you explain how you worked that out? Have you made any jottings?

Other things to do

- When the round is over, children check both their calculations on a calculator. If a child got theirs right, they score an extra point.

- For easier calculations, use 0–6 digit cards (0–7 digit cards for the last four problems).

Player 1

↑ ↓   43.6 + | 8 | 5 | . | 3 | = | 128.9 | 👍
                                                        👎

Player 2

↑ ↓   43.6 + | 7 | 6 | . | 1 | = | 119.7 | 👍
                                                        👎

ABOUT THE MATHS

In the teacher's session, children were encouraged to do additions and subtractions in the following way:

```
    ₁   ₁                    ⁸  ¹⁸
    4 3 . 6                  9̶ 8̶ . 9
+   2 9 . 7              -   3 9 . 2
  ─────────                ─────────
    7 3 . 3                  5 9 . 7
```

If they prefer to use another method, which they feel more confident about, they should do so.

Aim

↑ ↓ 43.6 + ☐☐.☐ = ☐ 👍 👎

↑ ↓ 52.8 + ☐☐.☐ = ☐ 👍 👎

↑ ↓ 73.5 + ☐☐.☐ = ☐ 👍 👎

↑ ↓ 98.9 + ☐☐.☐ = ☐ 👍 👎

↑ ↓ 25.2 + ☐☐.☐ = ☐ 👍 👎

↑ ↓ 16.47 + ☐☐.☐ = ☐ 👍 👎

↑ ↓ 99.53 + ☐☐.☐ = ☐ 👍 👎

↑ ↓ 36.72 + ☐☐.☐ = ☐ 👍 👎

↑ **Highest answer**   ↓ **Lowest answer**

Focus 20

Using a pencil-and-paper method for multiplication

Vocabulary

operation, calculation, calculate, multiply, product, estimate, approximately, check, method

Resources

RS20 TA for each child

0–9 dice for each pair

Scrap paper for working out

ACTIVITY

Pairs

Each child has their own copy of RS20 TA. Children take turns to roll the dice five times to get five digits. They write these down in the grid at the top of each round. One player decides whether, in this round, they are aiming for the highest or lowest answer, and each child circles the appropriate arrow on RS20 TA. Both children arrange their digits any way they like to make a four-digit number and a single-digit number. They write their two numbers in the boxes on the worksheet.

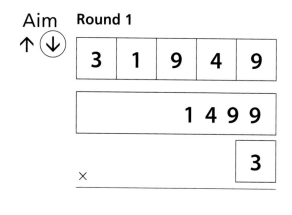

Each child then multiplies the four-digit number by the single-digit number and writes down the answer.

Children compare answers and agree the winner of that round (that is, the player with the highest or lowest answer).

Children continue for six rounds, taking turns to decide whether they are aiming for the highest or lowest answer. The child who wins most rounds is the overall winner.

How can you turn those numbers into the easiest possible calculation?

How are you going to do this calculation?

Explain to your partner how you did it.

Other things to do

- Talk with children about how to arrange their numbers to help them get a high or a low answer. Suggest that children use a 1–6 dice to make easier calculations.

- When the round is over, children check both their calculations on a calculator.

ABOUT THE MATHS

Children can choose how to do their multiplication. They are likely to use the grid method, which they used in the teaching session.

For example: **6239 × 5**

×	6000	200	30	9		
5	30 000	1000	150	45	=	31 195

Aim ↑ ↓   **Round 1**

×

Aim ↑ ↓   **Round 2**

×

Aim ↑ ↓   **Round 3**

×

Aim ↑ ↓   **Round 4**

×

Aim ↑ ↓   **Round 5**

×

Aim ↑ ↓   **Round 6**

×

↑ **Highest answer**   ↓ **Lowest answer**

Focus 21

Vocabulary

operation, divide, quotient, remainder, divisor, divisible by, estimate, approximately, check, decimal, decimal point

Resources

RS21 TA for each child

ACTIVITY

Individuals

Children choose a decimal number from RS21 TA and a one-digit number, then divide the decimal number by the one-digit number. They do six of these calculations.

Remind them to make an estimate first and to check their answers. Children show all their working.

What looks like the easiest pair of numbers to choose? Why?

Explain to your partner how you will split that number up to make the calculation easier.

Are there any numbers you can divide that you don't need to split into smaller parts?

Other things to do

- Help children check their divisions by doing a multiplication, with or without a calculator. For example, if they work out that $31.2 \div 6 = 5.2$, they can check this by multiplying 5.2 by 6.

- Children can invent easy division calculations with whole numbers, such as $55 \div 5$, and turn them into decimal calculations such as $5.5 \div 5$.

ABOUT THE MATHS

In the teacher's session, children used the following method for division. It involves splitting the larger number, which is too 'hard' to divide comfortably, into smaller parts that children feel confident they can divide. Children can choose to split the larger number however they like.

For example: $27.5 \div 5$

$27.5 = 25 + 2.5$

So

$27.5 \div 5 = (25 \div 5) + (2.5 \div 5)$

$= 5 + 0.5$

$= 5.5$

Encourage children to use each of the single-digit numbers on the sheet, rather than sticking with one number they feel safe with (probably the 2!).

Choose a decimal number and a one-digit number.
Divide the decimal number by the one-digit number. Do six calculations.

Remember to make an estimate first and to check your answer.
Show all your working.

4.8	28.8	45.6	19.2	33.6

2	3	4	6	8

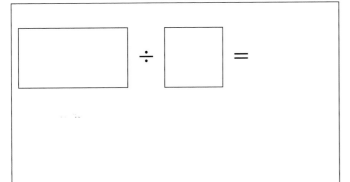

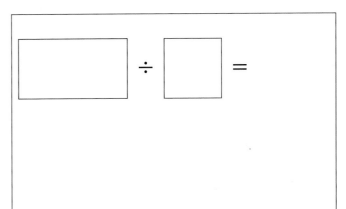

☐ ÷ ☐ =

☐ ÷ ☐ =

☐ ÷ ☐ =

☐ ÷ ☐ =

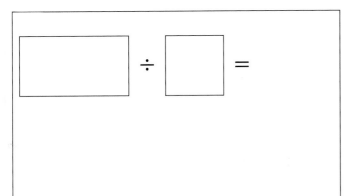

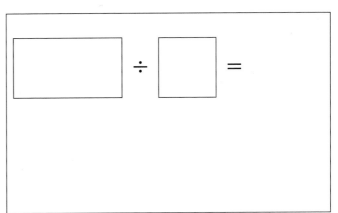

☐ ÷ ☐ =

☐ ÷ ☐ =

Focus 22

Vocabulary

operation, answer, method, strategy, explain, calculator, display, key, enter, clear, constant, memory

Resources

RS 22 TA for each pair
Calculator for each pair

ACTIVITY

Pairs, then groups of four

Children work together to solve each calculation, using a calculator. Child A has the calculator and decides what keys to press. They say out loud what these keys are, and Child B writes down the keys in order.

When the pair has finished the first calculation, they compare their work with another pair. They should check whether they have the same answers and whether they pressed the same keys, in the same order. If the pairs had different results for either of the above, they should try to establish which answer is correct.

Pairs then continue with the next calculation.

Let's try putting this calculation into the calculator in as many different ways as we can and find out all the different answers we can get. Which answers are right, and why?

Show your partner how you can use the memory function to hold on to that number before you do the next bit of the calculation.

How do you know that is the correct answer?

Other things to do

- Children show you how to use the memory function on the calculator. (It doesn't matter whether you already know, or not.)

- Children make up a calculation they then solve using the memory function.

ABOUT THE MATHS

Children have been learning to use the memory keys on a calculator. Encourage them to use the memory keys to help them with these calculations, but don't insist on it.

Answers to the calculations (remember: always do parts in brackets first):

1. $(56 \times 19) + 143 =$
   $56 \times 19 = 1064$
   $1064 + 143 = 1207$

2. $(7586 \times 6) - 217 =$
   $7586 \times 6 = 45516$
   $45516 - 217 = 45299$

3. $(19 \times 8) + (27 \times 14) =$
   $19 \times 8 = 152$
   $27 \times 14 = 378$
   $152 + 378 = 530$

4. $(54 \times 18) + (73 - 37) =$
   $54 \times 18 = 972$
   $73 - 37 = 36$
   $972 + 36 = 1008$

5. $(67 \times 32) + (196 \div 7) =$
   $67 \times 32 = 2144$
   $196 \div 7 = 28$
   $2144 + 28 = 2172$

6. $(414 \div 9) \times (405 \div 27) =$
   $414 \div 9 = 46$
   $405 \div 27 = 15$
   $46 \times 15 = 690$

1. $(56 \times 19) + 143 =$

2. $(7586 \times 6) - 217 =$

3. $(19 \times 8) + (27 \times 14) =$

4. $(54 \times 18) + (73 - 37) =$

5. $(67 \times 32) + (196 \div 7) =$

6. $(414 \div 9) \times (405 \div 27) =$

Focus 23

Vocabulary

parallel, perpendicular, edge, vertex/vertices, polygon, quadrilateral, parallelogram, rhombus, kite, isosceles, scalene, equilateral, diagonal, right-angled

Resources

RS23 TA for each child

Paper clip for each pair

Ruler for each child

ACTIVITY

Pairs

Each child has their own copy of RS23 TA. Children take turns to spin the spinner and read out what it says. (Hold the paper clip to the centre of the spinner with a pencil tip and flick it.)

If the spinner lands on 'perpendicular' (or 'parallel'), they find two dotted lines on the same shape that are perpendicular (or parallel) to each other. They use their ruler and draw over the two dotted lines.

The pair continue until both have drawn over all the lines on their sheet.

Explain why you think those lines are perpendicular to each other.

How could you show those lines are parallel?

Can you draw another line on that shape parallel to one edge?

Other things to do

- Children spin the spinner and read out what it says, then draw a shape with sides to match.

- You can make a game of this, working with the whole group in pairs. Each pair shares a sheet and has counters in two colours. You take charge of the spinner and read out what it says. The children whose turn it is find a shape with a pair of lines that fit your description and put one of their counters there. Continue like this.

ABOUT THE MATHS

Parallel lines go in the same direction and can never meet (even if they go on for ever). They can be of different lengths, start and end at different places and need not be horizontal.

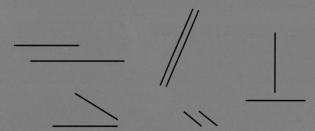

Perpendicular lines are at right angles to each other. They can be of different lengths, might, or might not, cross and might, or might not, touch.

perpendicular

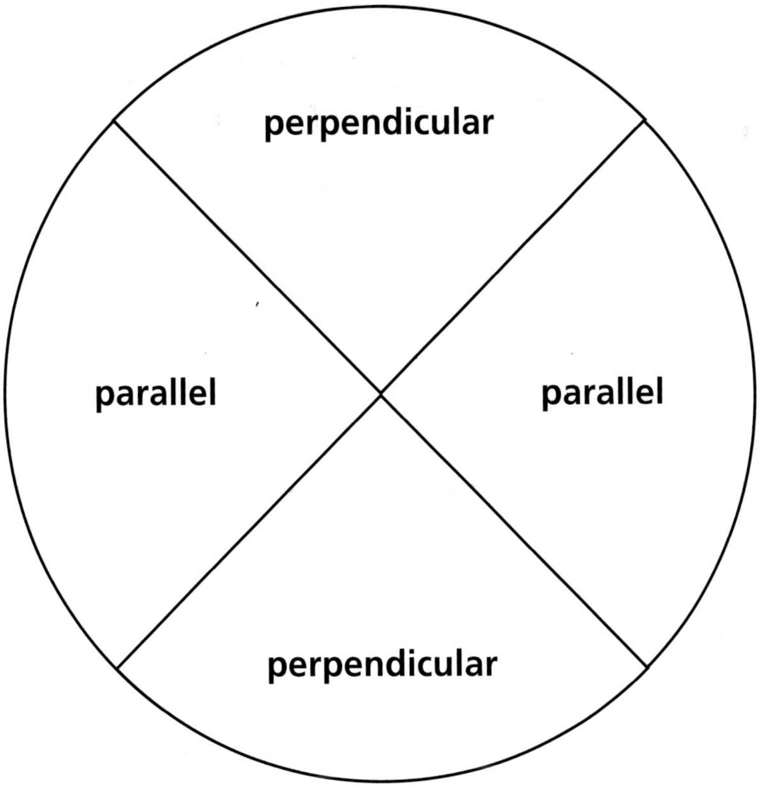

parallel

parallel

perpendicular

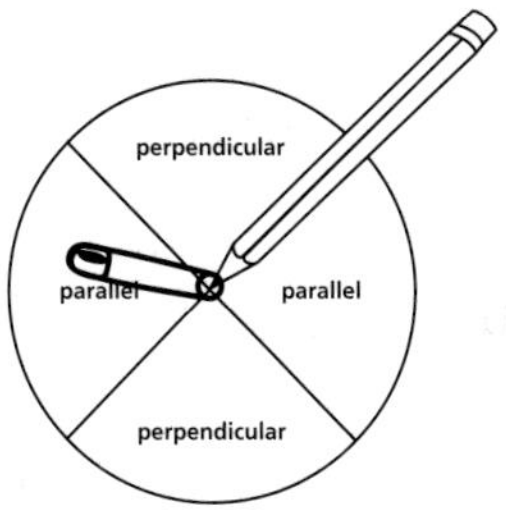

perpendicular

parallel

parallel

perpendicular

Focus 24

Vocabulary

position, direction, translate, translation, side, vertex, vertices, origin, coordinates, *x*-coordinate, *y*-coordinate, *x*-axis, *y*-axis, quadrant

Resources

RS24 TA (cut in half) for each child
Ruler for each child

ACTIVITY

Pairs

Each child writes their name on both of their grids. In secret, children translate each shape on one of their grids (that is, draw it again on the same grid, but in a different position). They can use a sideways or up-and-down movement or both, but they should *not* rotate the shape or reflect it. They can move each shape differently.

This is OK: These are not OK:

C C   **C Ɔ**   **C ∩**

Translation   Reflection   Rotation

They label the new shapes A1 to D1.

When they have finished, they describe how they moved one of their shapes, and their partner makes the same translation on the other unmarked grid. They then swap roles, so that both children have taken the role of instructor.

Translate shape A three squares to the right.

Can you describe how you have moved that shape?

How would you translate that shape back to its first position?

Other things to do

- Give instructions to the whole group on using one of the grids. For example say, "Translate shape A two squares to the right and two squares downwards." The children can then look at each other's work to check they have all done the same thing.

- Pairs of children sit back to back. One translates a shape on their grid, giving instructions to their partner as they do it. At the end, they check that the two grids are the same.

ABOUT THE MATHS

When a shape is translated, it is simply moved up, down or sideways through a combination of movements. As suggested above, a shape looks the same after translation, unlike with reflection and rotation.

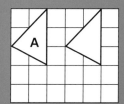

Shape A has been translated 3 squares to the right.

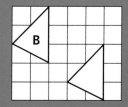

Shape B has been translated 3 squares to the right and 2 squares down.

Names: _____

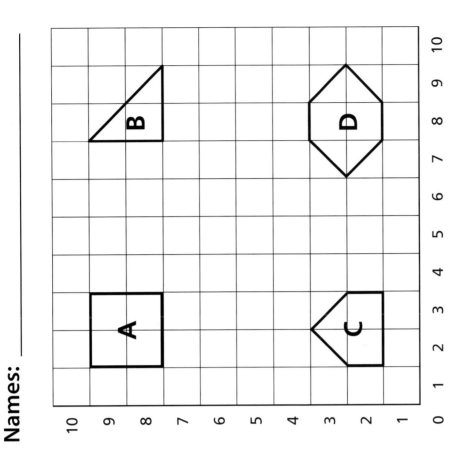

Names: _____

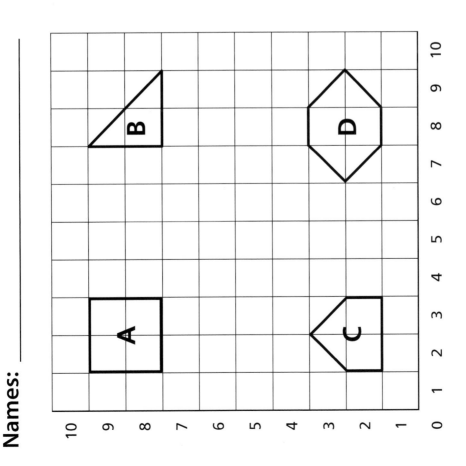

Focus 25

Measuring and estimating angles

Vocabulary

position, angle, acute, obtuse, right angle, protractor

Resources

RS25 TA for each child

Ruler for each child

Protractor for each child

ACTIVITY

Pairs

Each child has a copy of RS25 TA. They use a ruler to draw one triangle on the worksheet by joining any three dots and label the triangle ABC.

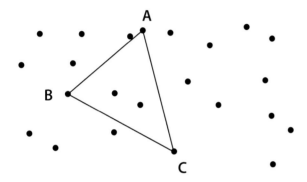

They then measure each of the angles and record them on the table.

The winner of the round is the child whose triangle contains the largest angle. The overall winner is the child who wins more rounds.

Which spots will give you the largest angle?

Where will you place the protractor to measure that angle?

How will you move the protractor to measure the next angle?

Other things to do

• Ask one of the children to tell you how they know which of the two scales on the protractor to read off when they measure an angle.

• Children estimate which angle of their triangle is the largest before they measure.

ABOUT THE MATHS

In the teacher's session, children learned how to use a protractor to measure angles such as B in the diagram below.

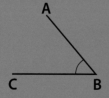

• Find the base line on the protractor (marked 0° at one end and 180° at the other).

• Place the centre of the base line exactly at the point of the angle B, with the base line running along the line CB.

• Hold the protractor firmly.

• To find which scale you need, check which one has its zero point on line CB.

• Look to see where the line AB meets that scale (in this example, it is the outer scale which shows the size of the angle).

• Read off the amount of angle at that point of the scale.

56 | **Mathematics Accomplished: Y6 Booster** Teaching assistant's book

Round	Angle A	Angle B	Angle C	Winner
1				
2				
3				
4				
5				

Focus 26

Converting metric units involving decimals

Vocabulary	Resources
estimate, measure, divide, multiply, convert, equivalent, unit	RS26 TA for each pair Pencils and paper

ACTIVITY

Whole group, then pairs

Read out the first four questions with the group. Take each problem in turn and ask the children to come up with ideas as to what they need to do to solve it (see *About the maths* below). Children make notes on their copy of RS26 TA.

Pairs of children then work together to answer the first four questions. Encourage the children to show all their working.

When children have answered the first four questions, they discuss as a pair how to answer the remaining two questions.

Would it help to underline all the words that describe measurements?

What is the relationship between grams and kilograms?

Can you show us all how you worked that out?

Other things to do

- Children research other sporting facts and write their own problems.

- Children rewrite one of the problems, using one of the original pieces of information and the answer to create a new problem.

For example:

Mike Powell jumped 5 cm short of 9 metres. How far did he jump?

ABOUT THE MATHS

The suggestions below are not the only methods children could use to solve the problems, but they are probably the easiest.

1. Knowing that 0.95 m is 5 cm short of a metre, 8.95 m is 5 cm short of 9 metres.

2. Find the difference between 54 kg and 57 kg (3 kg), then convert this to grams (3 kg = 3000 g).

3. Divide the number of centimetres by 100 to turn them into metres (760 cm = 7.6 m).

4. $\frac{1}{4}$ of 20 kg is 5 kg, so $\frac{3}{4}$ is 15 kg. Multiply the number of kilograms by 1000 to turn them into grams: 15 kg = 15 000 grams

5. Divide the 1500 m by the length of the pool: $1500 \div 50$

6. Add the total number of arrows and multiply by 10. The highest score is 720: $[(3 \times 12) + (3 \times 12)] \times 10$

7. 209 cm = 2.09 m

1. In Tokyo, the American athlete Mike Powell leapt 8.95 metres in the long jump. How many centimetres short of nine metres did he jump?

2. In boxing, a bantam weight boxer weighs 53.5 kg, and a featherweight boxer weighs 57 kg. How many grams heavier is a featherweight than a bantam weight?

3. In the game of rounders, the batting square must be 750 centimetres from the bowling square. How many metres is this?

4. Weightlifters hold up a barbell, which is a steel bar with weights on each end. The men's barbell weighs 20 kilograms. The women's barbell is three-quarters the weight of the men's. How many grams is that?

5. An Olympic-sized swimming pool is 50 m long and 25 m wide. How many lengths of the pool will a competitor in the 1500 m freestyle race have to swim?

6. In an archery competition, archers shoot 3 dozen arrows at 50 m and 3 dozen arrows at 40 m. An arrow hitting the bullseye scores 10. What is the maximum an archer can score in the competition?

7. When Stefka Kostadinova from Bulgaria did a high jump of 209 centimetres, the height was shown in metres on television. How many metres was that?

Focus 27

Vocabulary

scale, metric, unit, metre, centimetre, millimetre, kilogram, gram, litre, centilitre, millilitre, division, approximate

Resources

RS27 TA for each child

Coloured pencils for each child

Kitchen scales, bathroom scales, spring balance, and other weighing scales (optional)

ACTIVITY

Individuals, then pairs

Each child has a copy of RS27 TA. They mark the weights, as instructed, using coloured pencils. In pairs, they compare sheets and check each other's results. Challenge them to justify any differences in their recordings and clarify any misunderstandings. You could work together to mark up a final sheet together.

With the children, look at the dials on different weighing scales and identify the smallest units marked.

What's the heaviest weight you can measure on each of these scales?

What is the value of the smallest unit on each of these scales?

What sort of measuring scales have a circular dial like this?

Other things to do

- One child marks any value on one of the scales (for example, 3.1 kg), and their partner chooses another scale and marks the same value on that.

- Choose one of the scales and ask children to tell you the value of each unnumbered marker. For example, on the first scale, these go: 0.5 kg, 1.5 kg, 2.5 kg … (or 500 g, 1 kg 500 g, 2 kg 500 g …)

ABOUT THE MATHS

The key to this work is finding out the value of the intervals between each marker. Once children know this, marking the various weights becomes relatively straightforward.

Look at the range indicated on each scale and ask children to work out the value of the interim markers and the small markers.

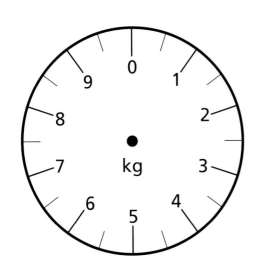

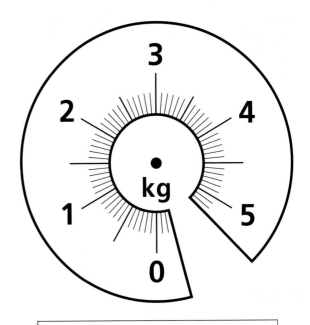

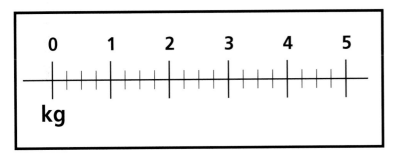

Mark these weights in the correct colours, on each of the three scales.

3.5 kg   red

$2\frac{1}{4}$ kg   blue

4200 g   green

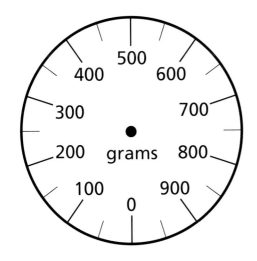

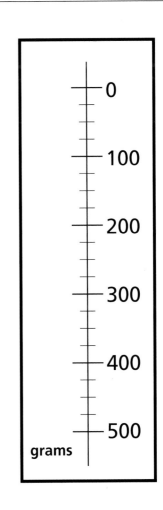

Mark these weights in the correct colours, on both scales.

50 g   red

0.5 kg   blue

0.1 kg   green

Focus 28

Vocabulary

measure, estimate, approximately, metric unit, standard unit, length, breadth, perimeter, area, surface, square centimetre (cm²), square metre (m²)

Resources

RS28 TA for each pair

Scissors for each pair

ACTIVITY

Pairs

Children cut out individual shapes (along the bold lines) from the grid on RS28 TA. They combine their shapes with a partner, so that they have two sets between them.

Child A chooses some of the pieces and puts them together. Child B says the area of the shape. Children then swap roles and repeat the activity.

Can you explain how you know the area is four squares?

I wonder what fraction of a square that triangle is?

Which of these shapes is worth just one square? What can you do to show us that it is?

Other things to do

- Child A says an area, and Child B makes a shape with that area.

- Child A says a perimeter, and Child B makes a shape with that perimeter.

ABOUT THE MATHS

Children work out the areas of their shapes by counting the squares. They match any half and quarter squares.

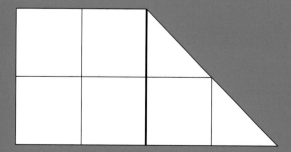

Area: 6 squares

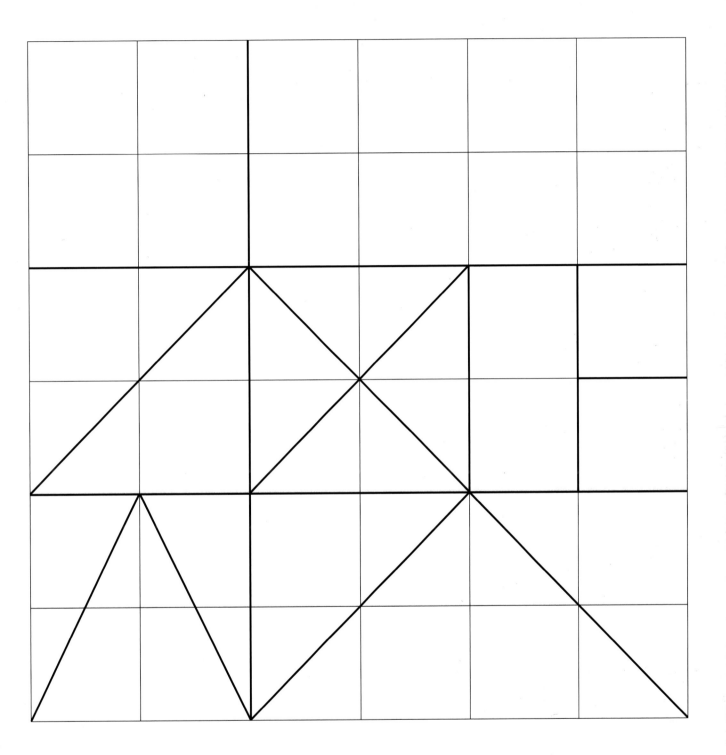

Focus 29

Estimating probability

Vocabulary

equal chance, even chance, fifty-fifty chance, equally likely, 1-in-2 chance, certain, random, impossible, possible, outcomes

Resources

RS29 TA for each pair
About 20 cubes in red, blue and black for each pair
Opaque bag for each pair

ACTIVITY

Pairs

Pairs of children choose one of the following probabilities and write it down on a slip of paper. They write their names on the paper to remind themselves of their choice.

- an even chance of getting red
- a poor chance of getting blue
- no chance of getting black

Each pair fills an opaque bag with 10 cubes to match their chosen probability. They swap bags with another pair *without* telling them which probability they chose. This other pair may not look inside the bag they are given. Instead, they experiment to find the intended probability by repeatedly:

- picking a cube from the bag and recording its colour in the tally chart on RS29 TA
- then putting the cube back in the bag and shaking the bag

After about 20, 50 or 100 cubes have been picked and replaced, children assess which of the three probabilities were chosen, recording this on the sheet. They check with the pair who filled the bag.

Can you explain to your partner what 'an even chance' means?

How could you change the contents of the bag to make a good chance of getting blue?

Do you think you'd get the same result if you did it another 50 times?

Other things to do

- Children invent three probabilities of their own and choose one in secret to use when filling their bag. They show the other pair what the three probabilities are, and this pair works out which one fits the bag contents.

ABOUT THE MATHS

Tallying uses the 'five-bar gate' marking system. For each cube you count, you draw a line. For the fifth cube, you cross off the previous four lines. Then you can easily count the final tally in fives.

Colour	Tally	Total
red	IIII IIII IIII IIII IIII II	27
Blue	IIII IIII IIII IIII II	22
Black	III	3

This table shows a likely 'even chance of getting red', even though red was chosen 27 times.

You would expect 'a poor chance of getting blue' to show just a few tally marks for blue and lots for the other colours.

For 'no chance of getting black' there would be no tallies at all for the black cubes.

The more times children repeat the experiment, the more accurate the probability will become.

Pick a cube from the bag and record its colour in the chart.

Put the cube back in the bag.

Shake the bag.

Now do the same thing again.

And again.

And again ... lots of times.

Colour	Tally	Total
red		
blue		
black		

Which probability does your bag represent? Tick the one you think is correct.

☐ an even chance of getting red

☐ a poor chance of getting blue

☐ no chance of getting black

Focus 30

Interpreting data

Vocabulary

data, represent, pie chart, bar chart, range, distribution, axis, mode, median, mean

Resources

RS30 TA for each pair

Shopping catalogues

Calculator for each pair (optional)

ACTIVITY

Pairs

Give each pair of children a page (or more) from a shopping catalogue showing a number of the same item such as toasters, watches, televisions …

Pairs record the prices shown on RS30 TA.

Expect at least five prices for each kind of item, although some prices may be the same. They then find the range, mode, median and mean of these prices, using a calculator as appropriate.

Children repeat this for a different item.

Do any of these T-shirts cost the same? Is there a mode in your list of prices?

What is the cheapest T-shirt and the most expensive? How does that help us find the range?

What is the mean average price of all these T-shirts? How would it change if you added one of these designer T-shirts in your list?

Other things to do

• Ask children to check each other's work.

• For the second kind of item on the worksheet, children choose an item and invent some prices. For example, they might invent prices for different computer games, DVDs, clothing or magazines.

ABOUT THE MATHS

Suppose a selection of toasters cost £9.99, £11.50, £12, £12, £12.90, £15.45, £23.95.

• The mode is to the most common or popular value (here, it is £12).

• The range is the difference between the lowest and highest values (£23.95 − £9.99, or £13.96).

• The mean is what we often think of as the average. It is calculated by adding all the values and dividing by the number of values (here, £97.79 ÷ 7 = £13.97).

• The median is the middle value when all the values are arranged in order. If there is no middle value, an imaginary value halfway between the middle two values is taken (£9.99, £11.50, £12, £12, £12.90, £15.45, £23.95; here, the median is £12, the 4th value).

Item			
Prices			
Mode	Range	Median	Mean

Item			
Prices			
Mode	Range	Median	Mean

×	1	2	3	4	5	6	7	8	9	10
1	1	2	3	4	5	6	7	8	9	10
2	2	4	6	8	10	12	14	16	18	20
3	3	6	9	12	15	18	21	24	27	30
4	4	8	12	16	20	24	28	32	36	40
5	5	10	15	20	25	30	35	40	45	50
6	6	12	18	24	30	36	42	48	54	60
7	7	14	21	28	35	42	49	56	63	70
8	8	16	24	32	40	48	56	64	72	80
9	9	18	27	36	45	54	63	72	81	90
10	10	20	30	40	50	60	70	80	90	100

1	2	3	4	5	6	7	8	9	10
11	12	13	14	15	16	17	18	19	20
21	22	23	24	25	26	27	28	29	30
31	32	33	34	35	36	37	38	39	40
41	42	43	44	45	46	47	48	49	50
51	52	53	54	55	56	57	58	59	60
61	62	63	64	65	66	67	68	69	70
71	72	73	74	75	76	77	78	79	80
81	82	83	84	85	86	87	88	89	90
91	92	93	94	95	96	97	98	99	100

1000	2000	3000	4000	5000	6000	7000	8000	9000
100	200	300	400	500	600	700	800	900
10	20	30	40	50	60	70	80	90
1	2	3	4	5	6	7	8	9
0.1	0.2	0.3	0.4	0.5	0.6	0.7	0.8	0.9
0.01	0.02	0.03	0.04	0.05	0.06	0.07	0.08	0.09

RS **Extra**

Kilograms and grams

1 kg	$\frac{3}{4}$ kg	$\frac{1}{100}$ kg	$\frac{1}{2}$ kg	$\frac{1}{10}$ kg	$\frac{1}{1000}$ kg	$\frac{1}{4}$ kg
1.0 kg	0.75 kg	0.01 kg	0.5 kg	0.1 kg	0.001 kg	0.25 kg
1000 g	750 g	10 g	500 g	100 g	1 g	250 g

Metres, centimetres and millimetres

1 m	$\frac{1}{4}$ m	$\frac{1}{1000}$ m	$\frac{1}{2}$ m	$\frac{3}{4}$ m	$\frac{1}{10}$ m	$\frac{1}{100}$ m
1.0 m	0.25 m	0.001 m	0.5 m	0.75 m	0.1 m	0.01 m
100 cm	25 cm	0.1 cm	50 cm	75 cm	10 cm	1 cm
1000 mm	250 mm	1 mm	500 mm	750 mm	100 mm	10 mm

Litres, centilitres and millilitres

1 l	$\frac{1}{100}$ l	$\frac{1}{1000}$ l	$\frac{1}{4}$ l	$\frac{1}{2}$ l	$\frac{3}{4}$ l	$\frac{1}{10}$ l
1.0 l	0.01 l	0.001 l	0.25 l	0.5 l	0.75 l	0.1 l
100 cl	1 cl	0.1 cl	25 cl	50 cl	75 cl	10 cl
1000 ml	10 ml	1 ml	250 ml	500 ml	750 ml	100 ml

Divisibility Rules

2	The last digit is 0, 2, 4, 6 or 8.
3	The sum of the digits is divisible by 3.
4	The last two digits are divisible by 4.
5	The last digit is 0 or 5.
6	The number is even and divisible by 3.
8	The last three digits are divisible by 8.
9	The sum of the digits is divisible by 9.
10	The last digit is 0.
25	The last two digits are 00, 25, 50 or 75.
100	The last two digits are 00.

Half	Number	Double
	1	2
1	2	4
	3	6
2	4	8
	5	10
3	6	12
	7	14
4	8	16
	9	18
5	10	20
	11	22
6	12	24
	13	26
7	14	28
	15	30
8	16	32
	17	34
9	18	36
	19	38
10	20	40
	21	42

Half	Number	Double
	10	20
10	20	40
	30	60
20	40	80
	50	100
30	60	120
	70	140
40	80	160
	90	180
50	100	200
	110	220
60	120	240
	130	260
70	140	280

Using a set square and ruler to draw perpendicular and parallel lines

Drawing perpendicular lines

Step 1: Draw a horizontal line with a ruler. Label the start of the line 'A' and the end of the line 'B'. Mark a point 'C'.	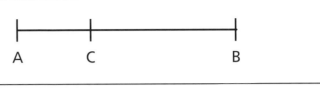
Step 2: Place the set square with its right-angled corner at point C. Make sure that the base of the set square lies exactly on the line AB. **Step 3:** Using the other perpendicular edge of the set square, draw the line CD.	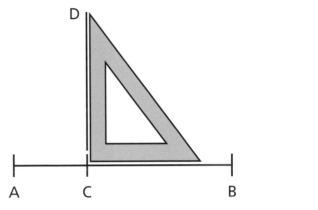

Drawing parallel lines

Step 1: Draw a horizontal line with the ruler. Label the start of the line 'A' and the end of the line 'B'. Mark a point 'C'.	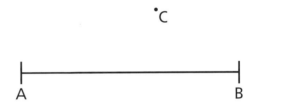
Step 2: Place the longest side of the set square on the line AB and the ruler against the shortest edge of the set square. Don't let the ruler be too close to point C. **Step 3:** Slide the set square along the ruler until the longest side of the set square touches point C.	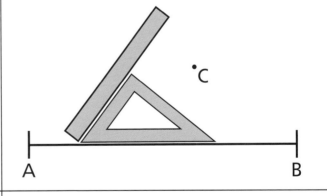
Step 4: Draw line CD along the longest side of the set square.	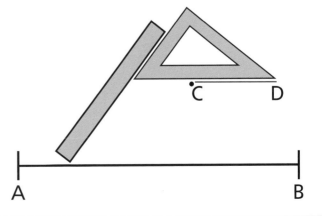

1 kilometre =	1000 metres
1 metre =	100 centimetres or 1000 millimetres
1 centimetre =	10 millimetres
1 kilogram =	1000 grams
1 litre =	1000 millilitres

kilometre	km
metre	m
centimetre	cm
millimetres	mm
kilogram	kg
grams	g
litre	l
millilitres	ml

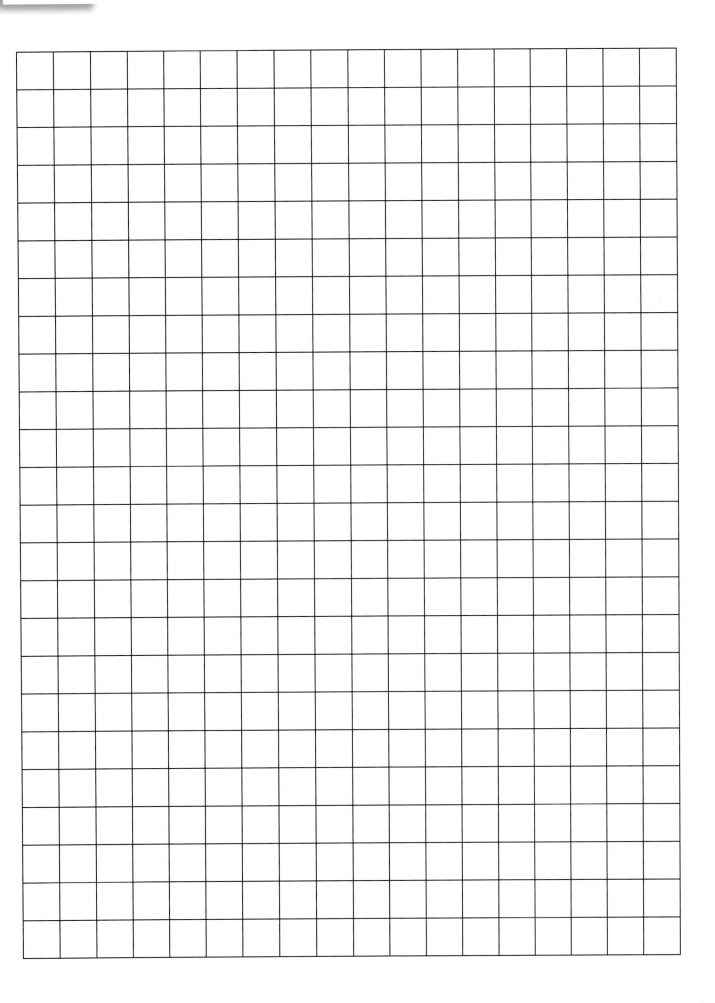

Visit us at www.beam.co.uk

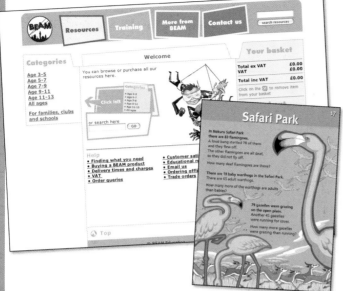

Browse for BEAM resources

You can search for resources, take a look
at sample pages and order online.

Try out some of our free activity sheets

You can download free games and activities for 3- to
13-year olds to use in your classroom or for homework.

Dip into some interesting research

You'll find some interesting research papers and
informative articles on maths education to download.

Take a look at our professional development

You can choose from our extensive range of courses,
or find out more about the BEAM Conference.

Sign up for the BEAM Bulletin

Our free monthly email newsletter will give you news on:

- Maths of the Month – a regular reminder, so you
  know when we've added new free downloadable
  games and activities

- New publications – a chance to find out more
  about our latest resources, including new titles
  not in the catalogue

- Courses and conferences – keep up to date with
  our PD programme, including new courses that
  we organise during the year

  To receive the BEAM Bulletin, email your name
  and email address to news@beam.co.uk